Towards a Theology
of the Environment

Towards a Theology
of the Environment

Paul Haffner

GRACEWING

First published in 2008
by

Gracewing
2, Southern Avenue
Leominster
Herefordshire
HR6 0QF
www.gracewing.co.uk

ISBN 978-0-85244-368-2

Cover design: view of Lake Tahoe, USA

Contents

Chapter 4: Christian Vision of Creation............183

Preface

This work is offered in devout homage to the Servant of God, Pope John Paul II, and to his vast, profound, and marvellous teaching concerning environmental issues. Since the beginning of his Pontificate, his successor Pope Benedict XVI has also described the essential core of a Christian ecology: 'External deserts are multiplying in the world, because interior deserts have become so vast. Therefore, the treasures of the earth are no longer serving the edification of the Lord's garden, in which all can live, but are rather subjected to the power of exploitation and destruction.'

I hope that this book can be an aid to students in this area, as well as to the lay people, clergy, and religious who wish to deepen their knowledge regarding the gift of creation which God has entrusted to us and which we experience in our Christian life. It is a general illustration of the mystery of creation in an environmental perspective, and it embraces the various Biblical, patristic, theological, moral and spiritual themes which intertwine in recent magisterial pronouncements on the topic.

The work begins by outlining some of the most serious problems that weigh on the environment today, and how they can be evaluated at the level of the relationships between man and the created world, in order to move on to a historical understanding of the ecological question. A distinction is drawn between the science of ecology and the ideological overtones often associated with this area. Next, an overview of Christian teaching on ecology is presented as an antidote to both New Age pseudo-mysticism and political ideology. The fourth chapter formulates a synthesis of environmental theology. This vision is then applied to some moral issues and to traditional spirituality. Everything is oriented toward the solution of various ecological problems.

This book attempts to provide a unique novelty, presenting an organic summary of environmental theology from the point of view of the Christian West and East. I hope that reading these pages can help all Christians deepen their understanding of environmental theology. In the preparation of this book I have been helped by my students: I thank them for their participation, which has greatly contributed to the formulation of the ideas found in this text.

I am most grateful to all those who have assisted in one way or another in the production of the English edition of this book. I wish to thank John Alexander Di Camillo for his help in preparing the text for publication, as well as for his suggestions and dedication in checking the drafts. Last, but by no means least, I am grateful as ever to Tom Longford, Director of Gracewing, for his help in producing this work.

Rome, 4 October 2008
Feast of Saint Francis of Assisi

Abbreviations

AAS=	*Acta Apostolicae Sedis.* Commentarium officiale. Typis Polyglottis Vaticanis, 1909-

CCC=	*Catechism of the Catholic Church.* LEV, Vatican City 1992.

DS=	H. Denzinger. *Enchiridion Symbolorum, Definitionum et Declarationum de rebus fidei et morum.* Edizione bilingue a cura di P. Hünermann. EDB, Bologna 1995.

IG=	*Insegnamenti di Giovanni Paolo II.* Vatican Polyglot Press, Città del Vaticano 1978-2005.

IP=	*Insegnamenti di Paolo VI.* Vatican Polyglot Press, Città del Vaticano 1963-1978.

OR=	*L'Osservatore Romano,* daily edition.

PA=

Papal Addresses to the Pontifical Academy of Sciences 1917-2002 and to the Pontifical Academy of Social Sciences 1994-2002. Vatican City: Pontifical Academy of Sciences, 2003

PG=

J.P. Migne. *Patrologiae cursus completus,* series graeca. 161 voll., Paris 1857-1866.

PL=

J.P. Migne. *Patrologiae cursus completus,* series latina. 221 voll., Paris 1844-1864.

1

The current situation

Today, many people think that the environmental problem concerning our planet is a 'clinical case'.[1] The resources of water, soil and air are often exploited to the extent that their capacity for renewal and self–purification is exceeded. For this reason, the variety of flora and fauna in their natural habitats is threatened, as is the amount of useable land for agriculture and recreational space. It is important to emphasize that the environmental situation is not entirely unfavourable. In recent years, through energetic measures in some countries, there has been considerable attention to and success in reducing the release of harmful substances into the water, air and soil. Unfortunately, in some areas, environmental damage spreads more rapidly than protection measures. We will now identify today's main environmental problems.

1.1 The nightmare of radiation

The disaster at the Chernobyl nuclear power plant in the Ukraine on 26 April 1986 was the greatest release of industrial atomic radiation in a single moment, causing all other episodes of less intense but more prolonged emissions, such as the episode at Three Mile Island (on 28 March 1979), to fade into the background. The problem of radioactive pollution, however, is only partly made up of these clamorous episodes: much more urgent is the safe conservation, for an estimated period of 25,000 years, of waste from power plants, industrial factories and major radiological centres. The ex–inhabitants of the Bikini atoll, over 50 years after the atomic tests carried out there, still cannot return. At this time the people of Bikini remain scattered throughout the Marshall Islands and the rest of the world as they wait for the cleanup of Bikini to begin in earnest. Another certain fact concerns the survivors of the bombings of Hiroshima and Nagasaki. In recent years, experts have greatly reduced what has been considered the maximum radiation exposure threshold, to the point that it should no longer be much greater than the natural background radiation level. This natural background radiation, however, varies according to geographic region.

1.1.1 Pylons and tumours

The oscillating electric and magnetic fields in electromagnetic radiation will induce an electric current in any conductor through which it passes. The best understood biological effect of electromagnetic fields is to cause dielectric heating. For example, touching an antenna while a transmitter is in operation can cause severe burns. This heating effect varies with the frequency of the electromagnetic energy. The eyes are particularly vulnerable to Radio Frequency (RF) energy in the microwave range, and prolonged exposure to microwaves can lead to cataracts. Each frequency in the electromagnetic spectrum is absorbed by living tissue at a different rate, called the specific absorption rate. Many national governments have established safety limits for exposure to various frequencies of electromagnetic energy based on the specific absorption rate.[2] Some publications support the existence of complex biological effects of weaker *non–thermal* electromagnetic fields, including weak Extremely Low Frequency (ELF) magnetic fields[3] and modulated Radio Frequency and microwave fields.[4] Fundamental mechanisms of the interaction between electromagnetic fields and biological material at non–thermal levels are not yet fully understood.[5] The definite existence and possible extent of non–thermal effects is not fully established.

In 1979, two researchers, Nancy Wertheimer and Ed Leeper, published an article based on their own epidemiologic study, alleging that the incidence of childhood leukaemia was higher in Denver neighbourhoods that were near electric power lines.[6] Their article generated a flurry of other studies. The idea was picked up by Paul Brodeur, who wrote a frightening three–part article for *The New Yorker* that reached a large and influential audience. Subsequent books by Brodeur in 1989 and 1993 alleged that power lines were 'Currents of Death' and that the power industry and the government were engaged in a cover–up.[7]

In 1990, three–year old Mallory Zuidema was suffering from Wilms' tumour, a rare kidney cancer. Her mother Michelle was tormented, as parents of stricken children must always be, by the question: 'Why my child?' She sought out Paul Brodeur, the investigative reporter who the year before had written of a connection between power–line fields and cancer for *The New Yorker*, a series that was then published as a book, *Currents of Death*. The cause of little Mallory's tumour, he suggested to Michelle, might well be the electromagnetic fields emanating from the nearby transmission lines of San Diego Gas & Electric Company. In May 1991, Ted and Michelle Zuidema sued the electric company, attributing their four–year–old daughter Mallory's tumour to high voltage wires that pass by just a few yards from the house. The girl was supposedly exposed to electromagnetic fields. The Zuidermas lost their case claiming that electricity sickened Mallory. Historically, this was one of the first legal cases directed against electric companies in virtue of a presumed cancer risk. Since then, electricity — a product in continual growth and the symbol itself of progress and well–being — has been accused of causing cancer (especially childhood leukaemia, lymphomas and brain tumours) through its electromagnetic radiation produced in razors and hair dryers, toasters and televisions, phones and stereo systems. Electrophobia includes essentially all of the electrical appliances that we use in the home and at work. However, there is a primary suspect, power lines, which up until this point had only raised questions of an aesthetic nature; now, they are perceived as a potential health threat on the order of incinerators, waste dumps, chemical factories and power plants that emit higher frequency waves such as radar, radio and television emitters and mobile phone antennae.

The term 'electromagnetic smog' has been coined, indicating a new risk factor that could combine with other forms of pollution, and could cause stress, insomnia and infertility. This feeds the nightmares of those who are in desperate search of a culprit for the various seemingly causeless illnesses. Nevertheless, it becomes quickly evident that the phenomenon of electromagnetic pollution leads to superstition and irrational fears. Perhaps these voices have a foundation, despite the fact that it still evades scientific certainty. Since 1979, the most important biomedical journals of the world (*American Journal of Epidemiology*, *The Lancet*) have begun to publish epidemiological and experimental studies hypothesizing the harmfulness of the magnetic fields emitted by alternating current electrical wires. The Scandinavian countries were the first to engage the issue, commissioning a series of epidemiological studies. After a long string of fruitless results, Anders Ahlbom, a Swedish researcher, evidenced a correlation between nearness to electrical wires and childhood leukaemia in 1992. According to this study, those who live within 100–150 metres of a large power line run a greater risk of contracting this illness than those who do not live near power lines. Nonetheless, the same Swedish researcher admits that the results of the study are based on a mere four cases above the average regarding the cancer in children. According to national statistics, in fact, out of the half of a million inhabitants taken into consideration, there were only 142 cases of leukaemia. Seen in this light, the risk constituted by power lines appears modest and not entirely proven. In 2001, Ahlbom and his team conducted a review into the impact of electromagnetic fields upon health, and found that there was a doubling in childhood leukaemia for magnetic fields of over 0.4 µTesla, though he summarised that 'this is difficult to interpret in the absence of a known mechanism or reproducible experimental support.'[8]

Ahlbom's findings were echoed by Draper and his team in 2005 when a 70% increase was found in childhood leukaemia for those living within 200 metres of an overhead transmission line, and a 23% increase for those living between 200 and 600 metres of the same line. Both of these results were statistically significant.[9] Bristol University (UK) has published work on a theory that could account for this increase, and would also provide a potential mechanism, namely that the electric fields around power lines attract aerosol pollutants.[10] There is still no final complete and convincing explanation. In a fact sheet produced by the World Health Organization in June 2007, the conclusion stated:

> Thus, on balance, the evidence related to childhood leukaemia is not strong enough to be considered causal. A number of other adverse health effects have been studied for possible association with ELF magnetic field exposure. These include other childhood cancers, cancers in adults, depression, suicide, cardiovascular disorders, reproductive dysfunction, developmental disorders, immunological modifications, neurobehavioral effects and neurodegenerative disease. The WHO Task Group concluded that scientific evidence supporting an association between ELF magnetic field exposure and all of these health effects is much weaker than for childhood leukaemia. In some instances (i.e. for cardiovascular disease or breast cancer) the evidence suggests that these fields do not cause them.[11]

What makes the existence of this risk even less certain is the lack of a convincing biological explanation of how electromagnetic fields might induce leukaemia. According to one of the most reliable hypotheses, this radiation acts on melatonin, a hormone secreted by the pineal gland, which seems to have an important role in fighting the development of tumours. In particular, the suggestion has been made that electromagnetic fields can reduce melatonin production, and thus increase the risk of breast cancer. However, yet again we find ourselves faced with hypotheses awaiting further experimental confirmation.[12]

1.1.2 Televisions and computer monitors

Computer monitors (CRT)

Behind the screen of a CRT (cathode ray tube) computer monitor, there lie various circuits and devices necessary for its functioning. Some parts of the monitor generate strong electromagnetic fields because they operate at very high voltages. These high frequency electromagnetic fields are liable to generate analogously harmful and perhaps even more damaging phenomena with respect those that generate so–called electromagnetic pollution. It is therefore necessary to use caution and attempt to respect several rules: the first is not to be too close to the screen; the further away, the better. It must be emphasized that almost all recently made monitors have precautionary devices in order to reduce this radiation as much as possible. A new, low–radiation monitor is therefore a good investment if one intends to sit in front of it for many hours each day. It is then good to avoid being next to or behind a monitor, because the magnetic field generated is much stronger in those positions. Furthermore, the process of electron deceleration in the CRT leads to X–ray emissions; currently, however, the design of the tube is such that it impedes the dispersion of these rays.

Computer monitors (LCD)

It is a whole different story for monitors with a liquid crystal display (LCD), given that they do not produce any electromagnetic pollution. The technology of active matrix TFT (thin film transistor) LCD monitors can boast comparable performance with those of traditional CRT monitors, but with the added benefit of no harmful radiation and flickering. Today there are TFT–LCD monitors not only for laptop computers, but also for home and office computers. Their price is constantly dropping, and by now they are taking the place of CRT monitors.

1.1.3 Mobile phones

Researchers at the Centenary Institute of Cancer Medicine and Cell Biology in Sydney, led by Tony Basten, have studied 200 mice that were genetically predisposed to develop lymphoma, a cancer of the white blood cells. Half of the mice were exposed to digital microwave pulses like those produced by common European mobile phones. Exposure to the radiation lasted thirty minutes, two times per day, with a maximum emission strength equal to that which a person undergoes with a mobile phone near his or her head. After 18 months, the mice exposed to the microwaves contracted from two to four times as many lymphomas than mice which were not exposed. After the researchers cleaned up the data, affected by a small number of kidney illness–related tumours discovered in some of the mice, they concluded that the exposed subjects still had double the number of lymphomas with respect to those not exposed. There were more cancers than expected. Experts in the effects of biological radiation are telling mobile phone users not to be alarmed, but they say that further studies must be conducted to estimate the risks. Further evidence comes from the Stewart Report which concluded by affirming that the radiation emitted by mobile phones affects the electrical activity of the human brain. However, it points out how epidemiological evidence does not suggest that exposure to this radiation produces cancer or damages other health aspects, reducing longevity.[13]

More recently, a study conducted by the Swedish Institute of Environmental Medicine concluded that the use of mobile phones for more than ten years increases the incidence of benign tumours in the acoustic nerve. The study examined the side of the brain on which the mobile phone is usually held, and discovered that the risk of an acoustic neuroma was four times higher with respect to the side where the device is not held. The study was conducted on 750 people, of whom 150 developed acoustic neuromas, and used only TACS phones (older generation), making it unreliable for GSM phones. Furthermore, it would have to be confirmed by other studies before reaching any conclusions. The neuromas are not lethal, but can grow to the point of becoming a mass that presses against the brain.

In addition, the hypothesis has been proposed that keeping a mobile phone in trouser pockets can reduce male fertility. According to a new study by the Hungarian University of Szeged, mobile phone use reduces sperm production in men by 30%. This is the first study that indicates that male fertility could be damaged by mobile phone emissions. Those at the highest risk are men who carry mobile phones in the appropriate case attached to the belt, or in their trouser pockets. Yet even spermatozoa which survive exposure would be partially damaged, further reducing fertility. Prolonged mobile phone use would therefore have negative effects both on spermatogenesis (sperm production) and on male fertility, insofar as lesser production would decrease spermatozoa concentration and mobility.

According to a study financed by the European Union, mobile phones could lead to a rare type of tumour in children. Mobile phone use could damage neurological development in children and pre–adolescents, compromising their academic results. The physical structure of children renders them particularly sensitive to radiation: the smaller dimensions of their cranium and reduced bone thickness in their heads allow for greater penetration and therefore greater absorption of radiation.

It would be significant to consider the connection between GSM mobile phone radiation and health problems in children such as headaches, sleep disturbances, memory reduction, nasal haemorrhages and greater frequency of epileptic seizures. A disquieting result emerges from the research: those who use these mobile phones have the greatest incidence of 'epithelial neuromas', rare tumours of the lateral peripheral area of the brain — the area that registers the highest levels of radiation penetration. A US Defence Intelligence Agency (DIA) document dated March 1976, reviewing Soviet work on biological effects of non–thermal exposure to microwave and radiofrequency radiation makes for interesting, but disturbing, reading. For not only have many of the effects reported now been found in the case of exposure to GSM telephony radiation, but the following extract (which, incidentally, was eventually also removed) reveals a lesser known 'dark side' of the issue that presaged — as it turned out — the subsequent deployment of this kind of radiation in psychotonics and other forms of non–lethal microwave weaponry:

> The potential for the development of a number of antipersonnel applications is suggested by the research published in the USSR, Eastern Europe and the West. Sounds and possibly even words which appear to be originating intracranially can be induced by signal modulation at very low average power densities. Combinations of frequencies and other signal characteristics to produce neurological effects may be feasible in several years. The possibility of inducing metabolic disorders is also suggested. Animal experiments reported in the open literature have demonstrated the use of low level microwave signals to produce death by heart seizure or by neurological pathologies resulting from breaching of the blood–brain barrier.[14]

According to more recent studies, the amount of electromagnetic radiation emitted by mobile phones damages the DNA of human cells in the laboratory. Researchers have discovered that radiation levels such as those coming from mobile phones evoke ruptures in the DNA strands of many human cells. This is a type of danger associated with tumours. Researchers have also found some hints, but no conclusive proof, of other cellular changes, including chromosomal damage, alterations in the activity of certain genes and an increase in the rate of cellular division. The dangerous effects occurred when the cells were exposed to an intensity of electromagnetic radiation between 0.3 and 2.0 watts per kilogram. A mobile phone usually emits an intensity of radiation between 0.2 and 1.0 watts per kilogram.[15]

1.1.4 Household appliances

In a study conducted in 1991 in Los Angeles, California, 232 cases of childhood leukaemia were examined in reference to just as many population controls, in an attempt to evaluate the claims that these cases had been caused by domestic exposure of the pregnant mother and the post–natal child to 60 Hz electric and magnetic fields.

The researchers took into consideration exposure to the following home appliances: electric blankets, waterbeds, analogue and digital alarm clocks, hair dryers, bedroom air conditioners, bedroom fans, bedroom convector heaters, black and white TVs, colour TVs, videogames, electric razors, electric curling irons, and microwave ovens. The scientific analysis, in its totality, provides evidence in favour of an association between exposure to 50/60Hz fields and childhood leukaemia. It is necessary, furthermore, to promote research aimed at the identification of the biological mechanisms underlying the possibly cancerous effects of 50/60 Hz magnetic fields, in order to provide an understanding of the currently available epidemiological results.

It is therefore necessary that plans for the assembly of new home appliances incorporate the goal of reducing exposure to electric and magnetic fields, including through the adoption of new technological solutions. In particular, exposure containment seems particularly important for day–care centres, schools and other environments intended for young children, whether indoors or outdoors. Analogous considerations would also be valid for the planning of other types of electrical devices. In all cases, the important factors are: (a) the frequency of the electric or magnetic field; (b) the distance from the source; and (c) the duration of exposure. Whatever the situation, the causes of tumours are complex.

Reduction of electric fields in the home environment can be achieved in various ways:

1) Installing a low voltage switch cabinet, which substitutes alternating voltage at 220V with a continuous low voltage at 9V so as to reduce the electric field.

2) Unplugging appliances that are not in use.

3) Not passing electrical wires behind the head of the bed and keeping electrical outlets away from the sides of the bed.

4) Not placing the bed against a wall with an electrical panel or large stationary electrical appliances on the other side (for example: washing machine, dishwasher, water heater).

5) Positioning alarm radios, clocks and bedside lamps which are powered by house wiring at least 50 cm away from the pillow when sleeping.

1.1.5 Airplane flights

The Earth and its inhabitants are being constantly showered by radiation from space. This steady shower of cosmic radiation is created by charged, sub–atomic particles (parts of atoms) that originate in our galaxy, other galaxies, and the sun. The particles interact with Earth's atmosphere and magnetic field to create cosmic radiation. The charged particles exhibit a wide range of energies and the rate at which cosmic rays bombard Earth depends on whether they are low– or high–energy. The various types of particles that constitute cosmic radiation includes neutrons, protons, electrons and photons. Neutrons have a capacity to cause cell damage 5 to 20 times greater than gamma and X–rays. Furthermore, they constitute a particularly strong type of radiation with regard to international flights, making up approximately 50% of the total dose of cosmic radiation. Some organs and tissues of the human body are particularly sensitive to the biological effects of radiation, for example the thyroid, bone marrow and lungs, and consequently it is important to be aware of the dose of this neutron radiation absorbed by these parts of the body. The vast majority of cosmic rays are low–energy. Although high–energy cosmic particles constantly pass through and sometimes interact with the body, they are very rare and very difficult to detect. About eight percent of our annual radiation exposure comes from outer space. The atmosphere shields us from cosmic radiation, and the more air that is between us and outer space, the more shielding we have. The closer we get to outer space, the more we are exposed to cosmic radiation. This holds true when we live at high altitudes or fly. The dose of cosmic radiation increases with increasing altitude, the length of the flight, and increasing latitude (getting closer to the North or South Pole) as well as solar activity.

Like radiation from other sources, cosmic radiation is measured in sieverts (Sv). Annual doses are measured in millisieverts (mSv) which are thousandths of a sievert. Measurements on an aircraft are measured in microsieverts (µSv) which are millionths of a sievert. All humans are exposed to background radiation at sea–level. This comes from sources such as the local environment, food and drink, medical exposure and building materials. The *per caput* annual dose equivalent at ground altitudes is estimated to be 270 µSv from charged particles and 50 µSv from neutrons. However, the doses received at flight altitudes are still considered very low. Most public travellers would not be exposed to more than 1 mSv per year.[16] A regular business traveller is likely to be exposed to more than 1 mSv per year, but assuming that they are undertaking the majority of their travel for business, they will come within the occupational exposure limits. Pregnant women should not be exposed to more than 1mSv per year. Pilots and flight crew may be exposed to approximately 3–4 mSv per year, and the yearly limit for occupational exposure on commercial air flights is 20 mSv per year. These figures are low when compared to Computerised Tomography (CT) scans of the chest (8 mSv) or abdomen (5–30 mSv). Since the crew in an airplane is clearly more exposed to cosmic radiation during flight activities, some international directives require airline companies to carry out periodic checks on the various routes.

1.2 Air pollution

Air pollution is the human introduction into the atmosphere of chemicals, particulates, or biological materials that cause harm or discomfort to humans or other living organisms, or damage the environment. Air pollution causes respiratory diseases and death. Air pollution is often identified with major stationary sources, but the greatest source of emissions is actually mobile sources, mainly automobiles. Gases such as carbon dioxide, which contribute to global warming, have recently gained recognition as pollutants by climate scientists, while they also recognize that carbon dioxide is essential for plant life through photosynthesis.

There are many substances in the air which may harm the health of humans, animals and plants. These arise both from natural processes and human activity. Substances not naturally found in the air or at greater concentrations or in different locations from usual are referred to as pollutants. Pollutants can be classified as either primary or secondary. Primary pollutants are substances directly emitted from a process, such as ash from a volcanic eruption or the carbon monoxide gas from a motor vehicle exhaust. Secondary pollutants are not emitted directly. Rather, they form in the air when primary pollutants react or interact. An important example of a secondary pollutant is ground level ozone — one of the many secondary pollutants that make up photochemical smog. However, some pollutants may be both primary and secondary: that is, they are both emitted directly and formed from other primary pollutants.

Major primary pollutants produced by human activity include sulphur oxides (especially sulphur dioxide) which are emitted from the burning of coal and oil. Nitrogen oxides (especially nitrogen dioxide) are produced in high temperature combustion, and are seen as the brown haze dome above cities or the plume downwind of cities. Carbon monoxide is a colourless, odourless, non–irritating but very poisonous gas; it is generated by incomplete combustion of fuel such as natural gas, coal or wood. Vehicular exhaust is a major source of carbon monoxide. Carbon dioxide is a gas emitted from combustion. Volatile organic compounds, such as hydrocarbon fuel vapours and solvents, are also emitted into the atmosphere. On another level, air pollution can be caused by particulate matter (PM), generally known as smoke and dust. PM10 is the classification of suspended particles 10 micrometres in diameter and smaller that will enter the nasal cavity.

PM2.5 has a maximum particle size of 2.5 micrometres and will enter the bronchial tubes and lungs. Toxic metals, such as lead, cadmium and copper are also atmospheric pollutants. Chlorofluorocarbons (CFCs), harmful to the ozone layer are emitted from products currently banned from use. Ammonia (NH_3) emitted from agricultural processes. Unpleasant odours can also pollute the air, and these can come from refuse, sewage and industrial processes. Radioactive pollutants can contaminate the air and are produced by nuclear explosions and war explosives, and natural processes such as radon. Secondary pollutants include particulate matter formed from gaseous primary pollutants and compounds in photochemical smog, such as nitrogen dioxide. Ground level ozone (O_3) is formed from nitrogen oxides and volatile organic compounds. Peroxyacetyl nitrate is similarly formed from nitrogen oxides and volatile organic compounds.

Sources of air pollution refer to the various locations, activities or factors which are responsible for the releasing of pollutants in the atmosphere. These sources can be classified into two major categories which are anthropogenic and natural sources. Anthropogenic sources (human activity) mostly relate to the burning of different kinds of fuel and include 'stationary sources' such as smoke stacks of power plants, manufacturing facilities, and municipal waste incinerators. Mobile sources like motor vehicles, aircraft, and marine vessels also contribute to air pollution. Similarly air pollution can simply be caused by burning wood or coal, and can emanate from fireplaces, stoves, furnaces and incinerators. Further sources are oil refining, and industrial activity in general, as well as chemicals, dust and controlled burn practices in agriculture and forestry management. To a certain extent the fumes from paint, hair spray, varnish, aerosol sprays and other solvents enter the atmosphere. Waste deposition in landfills generates methane which is an air pollutant. Military sources include nuclear weapons, toxic gases, germ warfare and rocketry. Natural sources of air pollution include dust from natural sources, usually large areas of land with little or no vegetation; also methane, emitted by the digestion of food by animals, for example cattle. Radon gas emitted from radioactive decay within the Earth's crust is a pollutant. Smoke and carbon monoxide from wildfires also enter the atmosphere. Volcanic activity can produce sulphur, chlorine, and ash particulates.

A lack of indoor ventilation concentrates air pollution where people often spend most of their time. Radon gas, a carcinogen, is exuded from the Earth in certain locations and trapped inside houses. Building materials including carpeting and plywood emit formaldehyde (H_2CO) gas. Paint and solvents give off volatile organic compounds as they dry. Lead paint can degenerate into dust and be inhaled. Intentional air pollution is introduced with the use of air fresheners, incense, and other scented items. Controlled wood fires in stoves and fireplaces can add significant amounts of smoke particulates into the air, inside and out. Indoor pollution fatalities may be caused by using pesticides and other chemical sprays indoors without proper ventilation. Carbon monoxide (CO) poisoning and fatalities are often caused by faulty vents and chimneys, or by the burning of charcoal indoors. Chronic carbon monoxide poisoning can result even from poorly adjusted pilot lights. Traps are built into all domestic plumbing to keep sewer gas, hydrogen sulphide, out of interiors. Clothing emits tetrachloroethylene, or other dry cleaning fluids, for days after dry cleaning.

Though its use has now been banned in many countries, the extensive use of asbestos in industrial and domestic environments in the past has left a potentially very dangerous material in many localities. Asbestosis is a chronic inflammatory medical condition affecting the tissue of the lungs. It occurs after long–term, heavy exposure to asbestos from asbestos–containing materials in structures. Sufferers have severe dyspnoea (shortness of breath) and are at an increased risk regarding several different types of lung cancer including mesothelioma, which is generally very rare; it is almost always associated with prolonged exposure to asbestos. Biological sources of air pollution are also found indoors, as gases and airborne particulates. Pets produce dander, people produce dust from minute skin flakes and decomposed hair, dust mites in bedding, carpeting and furniture produce enzymes and micrometre–sized faecal droppings, inhabitants emit methane, mould forms in walls and generates mycotoxins and spores, air conditioning systems can incubate Legionnaires' disease and mould, and houseplants, soil and surrounding gardens can produce pollen, dust, and mould. Indoors, the lack of air circulation allows these airborne pollutants to accumulate more than they would otherwise occur in nature. Smoking is a form of indoor air pollution which is now fortunately restricted by national legislation.

1.2.1 Acid rain

The term 'acid rain' is commonly used to indicate the deposition of acidic components in rain, snow, fog, dew, or dry particles. The more accurate term is 'acid precipitation.' Distilled water, which contains no carbon dioxide, has a neutral pH of 7. Liquids with a pH less than 7 are acidic, and those with a pH greater than 7 are bases. 'Clean' or unpolluted rain is slightly acidic, its pH being about 5.6, because carbon dioxide and water in the air react together to form carbonic acid, a weak acid. Carbonic acid then can ionize in water forming low concentrations of hydronium ions. The extra acidity in rain comes from the reaction of primary air pollutants, primarily sulphur oxides and nitrogen oxides, with water in the air to form strong acids (like sulphuric and nitric acid). The main sources of these pollutants are vehicles and industrial and power–generating plants.

Though acid rain was discovered in 1852, it wasn't until the late 1960's that scientists began widely observing and studying the phenomenon. Public awareness of acid rain in the U.S increased in the 1990's after the *New York Times* published reports from the Hubbard Brook Experimental Forest in New Hampshire of the myriad deleterious environmental effects demonstrated to result from it. Occasional pH readings of well below 2.4 (the acidity of vinegar) have been reported in industrialized areas. Industrial acid rain is a substantial problem in China, Eastern Europe, Russia and areas downwind from them, partly because these areas all burn sulphur–containing coal to generate heat and electricity.[17] The use of tall smokestacks to reduce local pollution has contributed to the spread of acid rain by releasing gases into regional atmospheric circulation. Often deposition occurs a considerable distance downwind of the emissions, with mountainous regions tending to receive the most (simply because of their higher rainfall). An example of this effect is the low pH of rain (compared to the local emissions) which falls in Scandinavia. Initially considered a problem limited to the Scandinavian regions, acid rain, snow and fog – caused by the encounter of precipitation with polluting gases, primarily from the use of carbon fossil fuels — have seriously damaged vegetation and compromised continental aquatic environments even in Canada, the USA, and Europe. As a consequence of atmospheric currents, there is often a border problem: pollutants are produced in countries different from the ones where the rains and other forms of precipitation actually fall. This renders solutions based on international agreements even more necessary.

Acid rain has been shown to have adverse impacts on forests, fresh waters and soils, killing off insect and aquatic life forms as well as causing damage to buildings and having possible impacts on human health. Research shows that not all fish, shellfish, or the insects that they eat can tolerate the same amount of acid; for example, frogs can tolerate water that is more acidic (namely has a lower pH) than trout. In the United States, many coal–burning power plants adopt *Flue gas desulphurisation* (FGD) to remove sulphur–containing gases from their stack gases. An example of FGD is the wet scrubber which is commonly used in the U.S. and many other countries. A wet scrubber is basically a reaction tower equipped with a fan that extracts hot smoke stack gases from a power plant into the tower. Lime or limestone in slurry form is also injected into the tower to mix with the stack gases and combine with the sulphur dioxide present. The calcium carbonate of the limestone produces pH–neutral calcium sulphate that is physically removed from the scrubber. That is, the scrubber turns sulphur pollution into industrial sulphates.

1.3 Waste

Urbanization and increasing industrialization have led to the problem of waste disposal. In some big cities, each inhabitant produces a daily average of around 2 kilograms of waste. While there is nearly a worldwide shortage of safe and appropriate locations for the millions of tons of toxic domestic and industrial waste, the dangers for environmental and human health are rapidly increasing. Our consumer society has inserted refuse into the programming of its production system, but without foreseeing the possible consequences with due attentiveness. At the same time, however, developing countries have begun to industrialize without acquiring the technology for safely treating dangerous chemical refuse. In fact, in some cases they have become 'refuse cans' for waste coming from 'rich' countries. At a more subtle level, the fabrication of products which the laws in more developed countries have prohibited — after having ascertained their harmfulness — is being continually channelled toward the Third World. The dangers are more serious in the case of cumulative poisons. Biodegradation is the process by which organic substances are broken down by the enzymes produced by living organisms. Biodegradability is an important issue in relation to waste disposal because it determines how long refuse will remain before decomposing.[18] It is also necessary to distinguish between toxic and non–toxic waste. A safe way of storing poisonous waste must be found, which requires a modification process.

Waste management methods vary widely between areas for many reasons, including type of waste material, nearby land uses, and the area available. On traditional method involves landfill. Disposing of waste in a landfill involves burying waste to dispose of it, and this remains a common practice in most countries. Historically, landfills were often established in disused quarries, mining voids or borrow pits. A properly–designed and well–managed landfill can be a hygienic and relatively inexpensive method of disposing of waste materials. Older, poorly–designed or poorly–managed landfills can create a number of adverse environmental impacts such as wind–blown litter, attraction of vermin, and generation of liquid leachate. Another common by–product of landfills is gas (mostly composed of methane and carbon dioxide), which is produced as organic waste breaks down anaerobically. This gas can create odour problems, kill surface vegetation, and contributes to the greenhouse effect. Design characteristics of a modern landfill include methods to contain leachate such as clay or plastic lining material. Deposited waste is normally compacted to increase its density and stability, and covered to prevent attracting vermin (such as mice or rats). Many landfills also have landfill gas extraction systems installed to extract the landfill gas. Gas is pumped out of the landfill using perforated pipes and flared off or burnt in a gas engine to generate electricity.

Incineration is a disposal method that involves combustion of solid, liquid and gaseous waste. Incineration and other high temperature waste treatment systems are sometimes described as 'thermal treatment'. Incinerators convert waste materials into heat, gas, steam, and ash, and the process is carried out both on a small scale by individuals, and on a large scale by industry. It is recognized as a practical method of disposing of certain hazardous waste materials (such as biological medical waste). However, incineration remains a controversial method of waste disposal, due to issues such as the emission of gaseous pollutants. Incineration is common in countries such as Japan where land is more scarce, as these facilities generally do not require as much area as landfills. Waste–to–energy (WtE) or energy–from–waste (EfW) are broad terms for facilities that burn waste in a furnace or boiler to generate heat, steam and electricity.

Recycling has become an effective method of refuse disposal and is simply the process of extracting resources or value from waste. There are a number of different methods by which waste material is recycled: the raw materials may be extracted and reprocessed, or the calorific content of the waste may be converted to electricity. New methods of recycling are being developed continuously, and are now described. The popular meaning of 'recycling' in most developed countries refers to the widespread collection and reuse of everyday waste materials such as empty beverage containers. These are collected and sorted into common types so that the raw materials from which the items are made can be reprocessed into new products. Material for recycling may be collected separately from general waste using dedicated bins and collection vehicles, or sorted directly from mixed waste streams. The most common recycled consumer products include aluminium beverage cans, steel food and aerosol cans, HDPE and PET bottles, glass bottles and jars, paperboard cartons, newspapers, magazines, and cardboard. Other types of plastic (PVC, LDPE, PP, and PS) are also recyclable, although these are not as commonly collected. These items are usually composed of a single type of material, making them relatively easy to recycle into new products. The recycling of complex products (such as computers and electronic equipment) is more difficult, due to the additional dismantling and separation required.

Table 1: Resin identification codes

Recycling No.	Abbreviation	Polymer Name	Uses once recycled
1	PETE or PET	Polyethylene terephthalate	Polyester fibres, thermoformed sheet, strapping, soft drink bottles.
2	HDPE	High density polyethylene	Bottles, grocery bags, recycling bins, agricultural pipe, base cups, car stops, playground equipment, and plastic lumber.
3	PVC or V	Polyvinyl chloride	Pipe, fencing, and non-food bottles.
4	LDPE	Low density polyethylene	Plastic bags, various containers, dispensing bottles, wash bottles, tubing, and various moulded laboratory equipment.
5	PP	Polypropylene	Auto parts and industrial fibres.

Recycling No.	Abbreviation	Polymer Name	Uses once recycled
	PS	Polystyrene	Desk accessories, cafeteria trays, toys, video cassettes and cases, insulation board and expanded polystyrene products (e.g. Styrofoam).
	OTHER	Other plastics, including acrylonitrile butadiene styrene acrylic, polycarbonate, polylactic acid, nylon and fibreglass.	

Waste materials that are organic in nature, such as plant material, food scraps, and paper products, can be recycled using biological composting and digestion processes to decompose the organic matter. The resulting organic material is then recycled as mulch or compost for agricultural or landscaping purposes. In addition, waste gas from the process (such as methane) can be captured and used for generating electricity. The intention of biological processing in waste management is to control and accelerate the natural process of decomposition of organic matter. There are a large variety of composting and digestion methods and technologies varying in complexity from simple home compost heaps, to industrial–scale enclosed–vessel digestion of mixed domestic waste. Techniques of biological decomposition are differentiated as being aerobic or anaerobic methods, though hybrids of the two methods also exist.

The energy content of waste products can be harnessed directly by using them as a direct combustion fuel, or indirectly by processing them into another type of fuel. Recycling through thermal treatment ranges from using waste as a fuel source for cooking or heating, to fuel for boilers to generate steam and electricity in a turbine. Pyrolysis and gasification are two related forms of thermal treatment where waste materials are heated to high temperatures with limited oxygen availability. The process typically occurs in a sealed vessel under high pressure. Pyrolysis of solid waste converts the material into solid, liquid and gas products. The liquid and gas can be burnt to produce energy or refined into other products. The solid residue (char) can be further refined into products such as activated carbon. Gasification is used to convert organic materials directly into a synthetic gas (syngas) composed of carbon monoxide and hydrogen. The gas is then burnt to produce electricity and steam.

An important method of waste management is the prevention of waste material being created in the beginning. Methods of avoidance include reuse of second–hand products, repairing broken items instead of buying new, designing products to be refillable or reusable (such as cotton instead of plastic shopping bags), encouraging consumers to avoid using disposable products (such as disposable cutlery), and designing products that use less material to achieve the same purpose (for example, light-weighting of beverage cans). In conclusion we should refer to the '3 Rs' of waste disposal: reduce, reuse and recycle, which describe waste management strategies according to their desirability in terms of waste minimization.

A particularly unpleasant form of waste is that which forms a type of 'plastic soup' floating in the Pacific Ocean and is growing at an alarming rate. The vast expanse of debris — in effect the world's largest rubbish dump — is held in place by swirling underwater currents. This drifting 'soup' stretches from about 500 nautical miles off the Californian coast, across the northern Pacific, past Hawaii and almost as far as Japan. The 'soup' consists of two linked areas, on either side of the islands of Hawaii, known as the Western and Eastern Pacific Garbage Patches. About one–fifth of the refuse — which includes everything from footballs and kayaks to Lego blocks and carrier bags — is thrown off ships or oil platforms. The rest comes from land. Plastic is believed to constitute 90 per cent of all rubbish floating in the oceans. Environmentalists fear that unless consumers cut back on their use of disposable plastics, the plastic stew would double in size over the next decade. In the past, rubbish that ended up in oceanic gyres has biodegraded. But modern plastics are so durable that objects half–a–century old have been found in the North Pacific dump. According to the UN Environment Programme, plastic debris causes the deaths of more than a million seabirds every year, as well as more than 100,000 marine mammals. Syringes, cigarette lighters and toothbrushes have been found inside the stomachs of dead seabirds, which mistake them for food. The slowly rotating mass of rubbish–laden water poses a risk to human health, too. Hundreds of millions of tiny plastic pellets, or nurdles — the raw materials for the plastic industry — are lost or spilled every year, working their way into the sea. These pollutants act as chemical sponges attracting man–made chemicals such as hydrocarbons and the pesticide DDT and then they enter the food chain.

1.4 The terrestrial environment

Many regions are damaged following the excessive use of fertilizers, the sedimentation of waste, and the presence of poisons. In many regions of the Earth, deserts are expanding as a consequence of the uncontrolled exploitation of grazing land and the erosion of cultivable terrain. In this damaging of our natural resources, specifically through deforestation and severe soil erosion, it is not only the poor farmer who has the greatest responsibility. There is a great lack of supervision, control, steady surveillance, and responsible and efficacious planning on the part of public authorities. There has been avarice and a lack of attention on the part of some landowners, who are incredibly insensitive to the ecological problem.

Soil contamination is the presence of man–made chemicals or other alteration in the natural soil environment. This type of contamination typically arises from the rupture of underground storage tanks, application of pesticides, percolation of contaminated surface water to subsurface strata, leaching of wastes from landfills or direct discharge of industrial wastes to the soil. The most common chemicals involved are petroleum hydrocarbons, solvents, pesticides, lead and other heavy metals. The occurrence of this phenomenon is correlated with the degree of industrialization and intensity of chemical usage. The concern over soil contamination stems primarily from health risks, both of direct contact and from secondary contamination of water supplies. Mapping of contaminated soil sites and the resulting cleanup are time consuming and expensive tasks, requiring extensive amounts of geology, hydrology, chemistry and computer modelling skills.

The major concern is that there are many sensitive land uses where people are in direct contact with soils such as residences, parks, schools and playgrounds. Other contact mechanisms include contamination of drinking water or inhalation of soil contaminants which have vaporized. There is a very large set of health consequences from exposure to soil contamination depending on pollutant type, pathway of attack and vulnerability of the exposed population. Chromium and obsolete pesticide formulations are carcinogenic to populations. Lead is especially hazardous to young children, in which group there is a high risk of developmental damage to the brain, while to all populations kidney damage is a risk. Chronic exposure to poisonous substances at sufficient concentrations is known to be associated with higher incidence of leukaemia. Obsolete pesticides such as mercury and cyclodienes are known to induce higher incidences of kidney damage, some irreversible; cyclodienes are linked to liver toxicity. Organophosphates and carbamates can induce a chain of responses leading to neuromuscular blockage. Many chlorinated solvents induce liver changes, kidney changes and depression of the central nervous system. There is an entire spectrum of further health effects such as headache, nausea, fatigue, eye irritation and skin rash for various chemicals.

Cleanup or remediation is analysed by environmental scientists who utilize field measurement of soil chemicals and also apply computer models for analysing the effect of soil chemicals. There are several principal strategies for remediation. First, it may be necessary to excavate the soil and remove it to a disposal site away from ready pathways for human or sensitive ecosystem contact. This technique also applies to the dredging of bay muds containing toxins.

Another possibility is the aeration of soils at the contaminated site, with the attendant risk of creating air pollution. A third approach is bioremediation, involving the microbial digestion of certain organic chemicals. Techniques used in bioremediation include landfarming, biostimulation and bioaugmentation of soil biota with commercially available microflora. Fourth, one can adopt the extraction of groundwater or soil vapour with an active electromechanical system, with subsequent stripping of the contaminants from the extract. Finally a solution could be the containment of the soil contaminants, such as by capping or paving over in place.

1.5 The disappearance of tropical rainforests

Tropical rainforests are generally found near the equator. They are common in Asia, Africa, South America, Central America, and on many of the Pacific Islands. Tropical and temperate rainforests have been subjected to heavy logging and agricultural clearance throughout the twentieth century, and the area they cover around the world is rapidly shrinking. Unless significant measures are taken on a world-wide basis to preserve them, by 2030 there may only be 10% of rainforest area remaining with another 10% in a degraded condition. 80% will have been lost and with them the natural diversity they contain could become extinct. Despite covering just 7% of the Earth's surface as of now, these forests are home to 50% of all known animal species and 80% of known plant species.[19] Tropical rainforests have been called the 'world's largest pharmacy', because of the large number of natural medicines discovered there. The disappearance of rainforest species eliminates potential uses, both in terms of genetic information for the improvement of cultivable plants and in terms of pharmacologically active compounds, which are instead destined to remain unknown. The causes can also be unexpectedly distant: the forests of Papua–New Guinea are turned into cardboard packaging for expensive electronic products. This indicates how buying furniture made of precious wood is not the only way that a citizen of a more developed country, thousands of kilometres away, can indirectly contribute to greater deforestation than a 'poor' person who lives there.

1.6 The aquatic environment

Over the years, oceans, lakes and rivers have become increasingly filled with harmful substances that slowly dissolve and converge into the food chain. The provision of drinkable water requires ever increasing technical expenses, which for a long time have no longer been sustainable by all countries and regions. Along with substances that are directly toxic, contaminants which destroy entire ecosystems by overfeeding them have gradually been added to the mix: this is called eutrophication, which kills by 'indigestion'. Eutrophication, strictly speaking, means an increase in chemical nutrients — typically compounds containing nitrogen or phosphorus — in an ecosystem. It may occur on land or in water. The term is however often used to mean the resultant increase in the ecosystem's primary productivity — in other words excessive plant growth and decay — and affects fish and other animal populations by depriving them of oxygen and severely reducing water quality. Eutrophication is frequently a result of nutrient pollution such as the release of sewage effluent and run–off from lawn fertilizers into natural waters (rivers or coasts), although it may also occur naturally in situations where nutrients accumulate (like in depositional environments) or where they flow into systems on an ephemeral basis (like with intermittent upwelling in coastal systems). Eutrophication generally promotes excessive plant growth and decay, favours certain weedy species over others, and is likely to cause severe reductions in water quality. In aquatic environments, enhanced growth of choking aquatic vegetation or phytoplankton (that is, an algal bloom) disrupts normal functioning of the ecosystem, causing a variety of problems such as a lack of oxygen in the water, needed for fish and shellfish to survive. This abnormal proliferation leads to algal collapse: first through the removal of oxygen, which other organisms need, and then through the emission of toxic decomposition gases. The water then becomes cloudy, coloured a shade of green, yellow, brown, or red. Human society is impacted as well: eutrophication decreases the resource value of rivers, lakes, and estuaries such that recreation, fishing, hunting, and aesthetic enjoyment are hindered. Health–related problems can occur where eutrophic conditions interfere with drinking water treatment. Eutrophication was recognized as a pollution problem in European and North American lakes and reservoirs in the mid–twentieth century.[20] Since then, it has become more widespread. Surveys showed that 54% of lakes in Asia are eutrophic; in Europe, 53%; in North America, 48%; in South America, 41%; and in Africa, 28%. Many people are aware of the episodes that occurred years ago on the Adriatic coast of Italy. The information from the artificial satellite Nimbus III has revealed a comparable situation along the coasts of Peru and Ecuador, where deep currents bring water up to the surface whose eutrophic content concentration comes from more highly industrial-ized areas. A solution has been identified in reducing the use of chemical fertilizers and increasing the use of organic ones.

Industries discharge a variety of pollutants in their wastewater including heavy metals, resin pellets, organic toxins, oils, nutrients, and solids. Discharges can also have thermal effects, especially those from power stations, and these too reduce the available oxygen. Silt–bearing runoff from many activities including construction sites, deforest-ation and agriculture can inhibit the penetration of sunlight through the water column, restricting photosyn-thesis and causing blanketing of the lake or river bed, in turn damaging ecological systems. Pollutants in water include a wide spectrum of chemicals, pathogens, and physical chemistry or sensory changes. Many of the chem-ical substances are toxic. Pathogens can produce waterborne diseases in either human or animal hosts. Alteration of water's physical chemistry include acidity, electrical conductivity, and temperature.

Water contaminants can be divided into organic and inorganic substances. Among organic water pollutants are insecticides and herbicides, with a huge range of organo-halide and other chemicals. There are also bacteria, which often come from sewage or livestock operations; food processing waste also includes pathogens. Pollution can also occur as a result of tree and brush debris from logging operations. Volatile organic compounds (VOCs), such as industrial solvents, can also pollute water if they are not stored properly. Dense non–aqueous phase liquids (DNAPLs), such as chlorinated solvents, may fall to the bottom of reservoirs, since they don't mix well with water and are more dense. Petroleum Hydrocarbons can cause water pollution problems and these include fuels (gasoline, diesel, jet fuels, and fuel oils) and lubricants (motor oil) from oil field operations, refineries, pipelines, retail service station underground storage tanks, and transfer operations. Detergents are harmful if they get into the water, as are various chemical compounds found in personal hygiene and cosmetic products. Similarly deleterious in large concentrations are disinfection by–products (DBPs) found in chemically disinfected drinking water. Inorganic water pollutants include heavy metals from various industrial processes, and acidity caused by industrial discharges (especially sulphur dioxide from power plants). Also troublesome are pre–production industrial raw resin pellets, chemical waste arising as industrial by–products, and silt in surface runoff from construction sites, logging, slash and burn practices or land clearing sites.

Thermal pollution is a temperature change in natural water bodies caused by human influence. The temperature change can be upwards or downwards. In the Northern Hemisphere, a common cause of thermal pollution is the use of water as a coolant, especially in power plants. Water used as a coolant is returned to the natural environment at a higher temperature. Increases in water temperature can affect aquatic organisms by decreasing oxygen supply, killing young fish which are vulnerable to small increases in temperature, and affecting ecosystem composition. In the Southern Hemisphere, thermal pollution is commonly caused by the release of very cold water from the base of reservoirs, with severe affects on fish (particularly eggs and larvae), and river productivity.

1.7 Ozone depletion

The important observation campaigns in 1986 and 1987 have shown that the meteorological peculiarities of the Antarctic regions in the winter and spring produce the specific conditions of an isolated air mass (the polar vortex) with temperatures low enough to justify the observed chemical disturbances. Numerous indications further confirm the conviction that chlorine compounds resulting from human activities are the primary culprits for the ozone reduction at the polar vortex. The chlorine primarily derives from the splitting of CFC (chlorofluorocarbon) molecules within the stratosphere, where they have been carried unaltered from their point of production on the Earth's surface. It is held that polar stratospheric clouds have a decisive role in the formation of the ozone hole above Antarctica. The satellite data demonstrate that the formation frequency of these clouds in the Antarctic stratosphere is greater than in any other area of the stratosphere itself. It was observed that in the spring, the reactive chlorine compounds are more than 50–100 times more abundant. An increase in the persistence of polar stratospheric clouds was also observed in 1985 and in 1987.

CFCs have been produced for various uses, and can be found in products such as solvents, foaming agents, refrigerator fluid and spray can propellant. The emission of chlorofluorocarbon gases, when combined with oxygen from the air, has led to a thinning of the ozonosphere. The photochemical decomposition of CFCs happens almost entirely in the stratosphere. The atmospheric layer of ozone normally filters the more biologically harmful components of ultraviolet rays. The 'ozone hole' has been seasonally revealed above Antarctica, sometimes with its thickness as much as halved. Consistent thinning has also been observed above the Arctic region, and, although still rarely at the moment, over the middle latitudes of the Northern hemisphere where the most densely populated countries are located. Automobiles and industries produce ozone, but it does not rise to the higher atmospheric levels: it remains below and contributes to lower level pollution.

In recent years, in the wake of studies during the 1970's which demonstrated that human activities can modify the total amount of atmospheric ozone and its vertical distribution, the number of studies on atmospheric ozone control processes has significantly increased. The fundamental reason for concern is the fact that ozone is the only gas in the atmosphere capable of preventing harmful ultraviolet solar radiation from reaching the planet's surface, and therefore of preventing the potentially damaging effects that an increase in such radiation could have on health (skin cancer, sight damage, immune system suppression) and on the productivity of aquatic and terrestrial ecosystems. Changes in the vertical distribution of ozone could alter the profile of atmospheric temperature and lead to climatic changes on a regional and global scale.

Atmospheric measurements are only available for the period extending back to 20 years ago, but the concentrations relative to previous years have been estimated with reasonable accuracy on the basis of production and emission data provided by the Chemical Manufacturing Association. Following a rapid increase of the most significant chlorofluorocarbon emissions — CFC–11 and CFC–12 — up through about 1970, there was a drop in the late 1970s due to restrictions on their use, which were introduced in some countries in light of the possible threat to the stratospheric ozone layer. From that point on, total

CFC use has continued to increase by about 4% per year, while aerosol propellants have decreased from 56% to 34% of total CFC production. At the beginning of the 1980s, the atmospheric concentrations of CFC–11 and CFC–12 were increasing by about 6% every year. In 1985, a Convention for the protection of the ozone layer was signed, which requires adhering countries to regulate CFC production. With the Montréal Protocol in 1987, a 50% reduction of CFC production by 1999 was agreed upon. At a meeting in London in June 1990, the same signers agreed on the elimination of CFC and carbon tetrachloride production by 2005. It was also established that hydrogenated CFCs (HCFCs) can be used in cases where no less dangerous alternative products are available, but that in any case these must also be eliminated by no later than 2040.

1.8 The Greenhouse Effect

The greenhouse effect is the process in which the emission of infrared radiation by the atmosphere warms a planet's surface. In addition to the Earth, Mars and Venus have greenhouse effects. The name comes from an incorrect analogy with the warming of air inside a greenhouse compared to the air outside the greenhouse. The name greenhouse effect is unfortunate, for a real greenhouse does not behave as the atmosphere does. The primary mechanism keeping the air warm in a real greenhouse is the suppression of convection (the exchange of air between the inside and outside). Thus, a real greenhouse acts like a blanket to prevent bubbles of warm air from being carried away from the surface. This is not how the atmosphere keeps the Earth's surface warm. Indeed, the atmosphere facilitates rather than suppresses convection. The greenhouse effect was discovered by Joseph Fourier in 1824 and first investigated quantitatively by Svante Arrhenius in 1896. There is a greenhouse effect, but, if there were not, we would all be dead! In the absence of the greenhouse effect, the Earth's average surface temperature of 14 °C would be about -19 °C. Global warming, a recent warming of the Earth's lower atmosphere, is believed to be the result of an enhanced greenhouse effect due to increased concentrations of greenhouse gases in the atmosphere.

The Earth receives energy from the Sun in the form of radiation. Most of the energy is in visible wavelengths and in infrared wavelengths that are near the visible range (often called 'near infrared'). The Earth reflects about 30% of the incoming solar radiation. The remaining 70% is absorbed, warming the land, atmosphere and oceans. For the Earth's temperature to be in steady state so that the Earth does not rapidly heat or cool, this absorbed solar radiation must be very closely balanced by energy radiated back to space in the infrared wavelengths. Since the intensity of infrared radiation increases with increasing temperature, one can think of the Earth's temperature as being determined by the infrared flux needed to balance the absorbed solar flux. The visible solar radiation mostly heats the surface, not the atmosphere, whereas most of the infrared radiation escaping to space is emitted from the upper atmosphere, not the surface. The infrared photons emitted by the surface are mostly absorbed in the atmosphere by greenhouse gases and clouds and do not escape directly to space.

The reason this warms the surface is most easily understood by starting with a simplified model of a purely radiative greenhouse effect that ignores energy transfer in the atmosphere by convection (sensible heat transport) and by the evaporation and condensation of water vapour (latent heat transport). In this purely radiative case, one can think of the atmosphere as emitting infrared radiation both upwards and downwards. The upward infrared flux emitted by the surface must balance not only the absorbed solar flux but also this downward infrared flux emitted by the atmosphere. The surface temperature will rise until it generates thermal radiation equivalent to the sum of the incoming solar and infrared radiation.

A more realistic picture taking into account the convective and latent heat fluxes is somewhat more complex. However the following simple model captures the essence. The starting point is to note that the opacity of the atmosphere to infrared radiation determines the height in the atmosphere from which most of the photons are emitted into space. If the atmosphere is more opaque, the typical photon escaping to space will be emitted from higher in the atmosphere, because one then has to go to higher altitudes to see out to space in the infrared. Since the emission of infrared radiation is a function of temperature, it is the temperature of the atmosphere at this emission level that is effectively determined by the requirement that the emitted flux balance the absorbed solar flux.

However, the temperature of the atmosphere generally decreases with height above the surface, at a rate of roughly 6.5 °C per kilometre on average, until one reaches the stratosphere 10–15 km above the surface. (Most infrared photons escaping to space are emitted by the troposphere, the region bounded by the surface and the stratosphere, so we can ignore the stratosphere in this simple picture.) A very simple model, but one that proves to be useful, involves the assumption that this temperature profile is simply fixed, by the non–radiative energy fluxes.

Given the temperature at the emission level of the infrared flux escaping to space, one then computes the surface temperature by increasing temperature at the rate of 6.5 °C per kilometre, the environmental lapse rate, until one reaches the surface. The more opaque the atmosphere, and the higher the emission level of the escaping infrared radiation, the warmer the surface, since one then needs to follow this lapse rate over a larger distance in the vertical. While less intuitive than the purely radiative greenhouse effect, this less familiar radiative–convective picture is the starting point for most discussions of the greenhouse effect in the climate modelling literature.

Atmospheric concentrations of the so–called 'greenhouse gases' have increased as a result of human activity. These gases are primarily carbon dioxide, methane, nitrous oxide, chlorofluorocarbons and tropospheric ozone (CO_2, CH_4, N_2O, CFC, O_3). These gases, even in small concentrations, can significantly alter the radiant equilibrium of the earth–atmosphere system. It is said that recent years have been the hottest in history, and that upcoming years will be even more so if atmospheric emissions of these gases deriving from human activities, especially carbon dioxide, are not significantly reduced. These gases increase the capacity of the atmosphere to trap the heat of solar radiation.

1.8.1 Carbon dioxide

Carbon dioxide, of which there is twice as much as all the other gases combined, is released by automobiles, industries, thermoelectric power plants and, in the Third World, by deforestation fires. In short, carbon dioxide is released into the atmosphere. Solar radiation is absorbed by the earth. The earth's surface in turn releases heat. The heat then remains partially trapped by the excess of carbon dioxide. If the current rate of progress is not halted, it is estimated that in less than a human lifespan (about 60 years), by 2050, this effect will have at least doubled, raising the global temperature of the Earth by 4°C, making the temperate zones hotter and dryer, and the desert and tropical zones extremely precipitous. While spontaneous flora and fauna could slowly adapt, crops and farm animal breeding would be much more easily disrupted. The greatest temperature increase will occur in the polar zones, whose increasingly melting icecaps will raise water levels.

The total average temperature of the earth's surface can be a useful indicator of the presence of effects due to changing concentrations of greenhouse gases. For many reasons, however, more detailed forecasts of regional temperature changes, precipitation amounts and other climatic parameters are necessary. As a consequence of the limits of climate models, the uncertainty in regional climatic change forecasts is greater than in those of global temperature changes. The degree of reliability of the estimates of global temperature changes, sea level, precipitation and evapotranspiration is relatively high; in contrast, the estimates of regional changes in these and other climatic variables have medium to low levels of certainty. The modifications of the variability of all of these elements are uncertain from one year to the next.

1.8.2 Methane

In recent years, there has been an annual average increase of about 1.1% in the concentration of methane in the atmosphere in the middle latitudes of the Northern Hemisphere. Farther–reaching trends can be obtained from an analysis of the air bubbles trapped in the ice layers in Greenland

and Antarctica. Methane is produced by microbial activities that occur during the mineralization of organic carbon in rigorously anaerobic conditions, such as in marshy soils and the intestines of herbivores. There is a considerable amount of uncertainty concerning the contributions of natural and anthropic sources of methane, and only order of magnitude estimates are possible.

1.8.3 Nitrous oxide

A recent increase in the concentration of nitrous oxide, or dinitrogen monoxide (N_2O), has been observed in the troposphere, despite having a considerably lower rate of increase than methane. The observations carried out between 1976 and 1980 showed an annual increase of 0.2–0.3%. Nitrous oxide has been reduced in the stratosphere, but no significant ways are known for eliminating it in the troposphere. Its emission into the atmosphere is primarily due to microbial activity in soil and water, and it enters into the nitrogen cycle. A considerable amount of uncertainty remains concerning emissions due to human activities, particularly with regard to the levels of nitrous oxide deriving from the combustion of biomasses and cultivated soil, especially involving the intensive use of fertilizers.

1.8.4 Tropospheric ozone

While stratospheric ozone has been reduced, tropospheric ozone has increased. The tropospheric ozone concentration is augmenting due to photochemical processes in which methane, carbon monoxide, hydrocarbons and nitrogen oxides carry out an important role. The concentration increase of these compounds, and their reactions with the hydroxyl radical (OH), are important factors for the chemistry of ozone in the troposphere. It is difficult to evaluate the changes on a planetary scale because of variations in the atmospheric ozone concentration from one zone to another. However, a net increase has been recorded in the middle and high latitudes of the Northern Hemisphere over the past 20–30 years, in particular during the summer months. The current rate of change is estimated to be about 1–2% per year.

1.9 Climate change

The term 'climate change' is adopted for human–caused change, and 'climate variability' for other changes. The term 'global warming' is a specific example of climate change, which can also refer to global cooling. In common usage, global warming often refers to the warming that can occur as a result of increased emissions of greenhouse gases from human activities. The term 'anthropogenic global warming' is also sometimes used when focusing on human–induced changes. The effects of forcing agents on the climate are complicated by various feedback processes. One of the most pronounced feedback effects relates to the evaporation of water. Warming by the addition of long–lived greenhouse gases such as carbon dioxide will cause more water to be evaporated into the atmosphere. Since water vapour itself acts as a greenhouse gas, the atmosphere warms further; this warming causes more water vapour to be evaporated, and so on until a new dynamic equilibrium concentration of water vapour is reached with a much larger greenhouse effect than that due to carbon dioxide alone. Although this feedback process causes an increase in the absolute moisture content of the air, the relative humidity stays nearly constant or even decreases slightly because the air is warmer.[21] This feedback effect can only be reversed slowly as carbon dioxide has a long average atmospheric lifetime. Feedback effects due to clouds are an area of ongoing research. Seen from below, clouds emit infrared radiation back to the surface, and so exert a warming effect; seen from above, clouds reflect sunlight and emit infrared radiation to space, and so exert a cooling effect. Whether the net effect is warming or cooling depends on details such as the type and altitude of the cloud. These details are difficult to represent in climate models, in part because clouds are much smaller than the spacing between points on the computational grids of climate models.

Global warming is the increase in the average temperature of the Earth's near–surface air and oceans in recent decades and its projected continuation. The global average air temperature near the Earth's surface rose 0.74 ± 0.18 °C during the 100 years ending in 2005.[22] The Intergovernmental Panel on Climate Change (IPCC) concludes that most of the observed increase in globally averaged temperatures since the mid–twentieth century is very likely due to the observed increase in anthropogenic greenhouse gas concentrations via the greenhouse effect.[23] Natural phenomena such as solar variation combined with volcanoes probably had a small warming effect from pre–industrial times to 1950 and a small cooling effect from 1950 onward.[24] Some scientists propose that increasing global temperature will cause the sea level to rise, and is expected to increase the intensity of extreme weather events and to change the amount and pattern of precipitation. Other effects of global warming include changes in agricultural yields, trade routes, glacier retreat, species extinctions and increases in the ranges of disease vectors.

The scientific community has always been divided as to the causes of the temperature increase recorded over the last century. Various experts consider it part of the normal alternation between cold and warm periods that has accompanied the millions of years of life existing on Earth. The question also arises as to whether the sun, with its more intense energetic activity, might not be an important contributor to global warming. There are other researchers who affirm that the warming process is accelerating due to the exponential increase of greenhouse gases. The temperature increase by decade has passed from 0.06°C (1905–1975) to 0.13°C (1975–1995) to 0.25°C (1995–2005). This spike is reflected in the increase in ocean temperatures, which are comparable to bathtub temperatures in the summer, melting icecaps, and rising water levels. The Arctic region loses 3% of its ice surface every ten years, while the ocean water level now rises at an average rate of 3 centimetres per decade, with points that are up to 10 times that rate in some areas, such as the Mauritius Islands (Indian Ocean) and Tuvalu (Pacific Ocean).[25]

Nevertheless, recent observations of phenomena such as glacial retreats, sea–level rise and the migration of temperature–sensitive species are not evidence for abnormal climate change, for none of these changes has been shown conclusively to lie outside the bounds of known natural variability. Moreover, the average rate of warming of 0.1 to 0.2 degrees Celsius per decade recorded by satellites during the late twentieth century falls within known natural rates of warming and cooling over the last 10,000 years. Leading scientists, including some senior IPCC representatives, acknowledge that today's computer models cannot predict climate. Consistent with this, and despite computer projections of temperature rises, there has been no net global warming since 1998. That the current temperature plateau follows a late twentieth century period of warming is consistent with the continuation today of natural multi–decadal or millennial climate cycling.

Scientists at the University of Rochester and the University of Virginia have published the results of two new studies: they cast doubts on global warming and on the validity of the climate models currently in use. Both of the studies were published in Geophysical Research Letters on July 9, 2004. The authors of the studies are David Douglass of the University of Rochester, Fred Singer of the University of Virginia and President of the Science and Environmental Policy Project (SEPP), Paul Knappenberger of New Hope Environmental Services, and Patrick Michaels of the University of Virginia.

The first study, titled 'Altitude dependence of atmospheric temperature trends: climate models versus observations', examines the well–known disparity between thermometric measurements of the soil, which show a warming trend, and those carried out by balloons or satellites in the lower atmosphere, which do not show any significant warming. The study analyses a method of measuring temperature in which historical meteorological data are used to determine the temperature values for each grid point of the earth, at an altitude of two metres. Yet, with this methodology, the data at the surface and those of the balloons or satellites coincide, showing that the climate is not warming.[26]

The second study, 'Disparity of tropospheric and surface temperature trends: new evidence', deals with another important issue: the disparity between global climate models and the data of the past quarter century.

All of these elaborate and costly computerized models agree on the fact that the introduction of greenhouse gases, such as CO_2, could lead to an increase in soil temperature. The latter would be directly proportional to altitude, doubling at three miles of altitude. The scientists at Rochester and Virginia, comparing the results of three of the most common climate models with four independent groups of observational data, have found that, while all of the models predicted a positive direction in the temperature trend, the observations showed negative values. Professor Michaels commented on the results of the study by explaining that it is not a surprise that the climate models, which make mistakes even in determining the soil temperature, cannot predict at all the effect of altitude. The surprising thing is that serious scientists seriously consider the results of these models.[27]

Some would argue that it is not possible to stop climate change, a natural phenomenon that has affected humanity through the ages. Doubt has been cast about the practicability of reducing carbon dioxide emissions:

> Without a doubt, atmospheric carbon dioxide is increasing due primarily to carbon–based energy production (with its undisputed benefits to humanity) and many people ardently believe we must 'do something' about its alleged consequence, global warming. This might seem like a legitimate concern given the potential disasters that are announced almost daily, so I've looked at a couple of ways in which humans might reduce CO_2 emissions and their impact on temperatures. California and some Northeastern states have decided to force their residents to buy cars that average 43 miles–per–gallon within the next decade. Even if you applied this law to the entire world, the net effect would reduce

projected warming by about 0.05 degrees Fahrenheit by 2100, an amount so minuscule as to be undetectable. Global temperatures vary more than that from day to day. Suppose you are very serious about making a dent in carbon emissions and could replace about 10% of the world's energy sources with non–CO_2–emitting nuclear power by 2020 — roughly equivalent to halving U.S. emissions. Based on IPCC–like projections, the required 1,000 new nuclear power plants would slow the warming by about 0.2 degrees Fahrenheit per century. It's a dent. But what is the economic and human price, and what is it worth given the scientific uncertainty?[28]

Geological, archaeological, oral and written histories all attest to the dramatic challenges posed to past societies from unanticipated changes in temperature, precipitation, winds and other climatic variables. It is therefore necessary to adapt, which the human person is capable of doing: the need is to equip nations to become resilient to the full range of these natural phenomena by promoting economic growth and wealth generation.

The Holy See has tended to steer a middle course between those who hold the idea of a human–induced climate change and those who propose a natural climate–swing theory:

Climate change is a serious concern and an inescapable responsibility for scientists and other experts, political and governmental leaders, local administrators and international organizations, as well as every sector of human society and each human person. My delegation wishes to stress the underlying moral imperative that all, without exception, have a grave responsibility to protect the environment. Beyond the various reactions to and interpretations of the reports of the Intergovernmental Panel on Climate Change (IPCC), the best scientific assessments available have established a link between human activity and climate change. However, the results of these scientific assessments, and the remaining uncertainties, should neither be exaggerated nor minimized in the name of politics, ideologies or self–interest. Rather they now need to be studied closely in order to give a sound basis for raising awareness and making effective policy decisions.[29]

In any case, the conventional viewpoint on global warming is suspect on a number of grounds. First, global warming is not the devastating threat to the planet it is widely alleged to be; moreover the remedy that is currently being proposed, which is in any event politically unattainable, would be worse that the threat it is supposed to avert.[30]

1.10 Automobile pollution

In the past, there was a problem of lead release deriving from tetra–ethyl lead used as an antiknock agent in gasoline. This antiknock agent is now prohibited in almost all countries. Nevertheless road transport accounts for 22% of total UK emissions of carbon dioxide, a major greenhouse gas. The European Union has voluntary agreements with motor manufacturers that aim to reduce average carbon dioxide emissions from new cars. Colour–coded labels, similar to those used on washing machines and fridges, are now displayed in car showrooms showing how much carbon dioxide new models emit per kilometre. However, as traffic levels are predicted to increase, road transport will continue to be a significant contributor to greenhouse gas emissions. Other air pollutants from road transport include nitrogen oxides, particles, carbon monoxide and hydrocarbons. Carbon monoxide and other combustion gases (primarily sulphur and nitrogen oxides), in addition to so–called 'particulates' or 'fine particles,' are still the principal problem with the use of internal combustion engines. An estimated 25–30% of all pollution is attributed to them. All have a damaging impact on the health of people, animals and vegetation locally. In town centres and alongside busy roads, vehicles are responsible for most local pollution. Vehicles tend to emit more pollution during the first few miles of journey when their engines are warming up. Although new technology and cleaner fuel formulations will continue to cut emissions of pollutants, the increasing number of vehicles on the road and miles driven is eroding these benefits. Noise from road traffic affects a significant proportion of people in industrialized countries. Sources include engine noise, tyre noise, car horns, car stereos, door slamming, and squeaking brakes. The sound of engines is a problem in towns and cities, while in more rural areas tyre noise on busy roads, which increases with speed, is the main source. Low–noise road surfaces, effective noise barriers in sensitive locations, and low noise tyres can all help reduce noise levels. Meanwhile, encouraging people to close car windows when playing loud music, and discouraging the use of 'boom box' car stereos would significantly reduce noise impact. Furthermore, a negative impact on the environment also derives from the disposal of the frames of discarded automobiles. New technologies that utilize electric, hybrid, or hydrogen combustion engines, and in a more distant future even 'fuel cells,' could eliminate or greatly reduce these problems.

1.11 Synthetic poisons

An example is DDT (dichloro–diphenyl–trichloroethane), the most well–known of certain chlorinated hydrocarbons, which is capable of eliminating all of the parasites that torment humanity and contribute to the spread of illnesses. Such is the case of lice that carry bubonic plague, mosquitoes that spread malaria and yellow fever, and so forth. The new poisons seemed entirely beneficial. Then in the 1950's it was confirmed that substances like DDT have the capacity — already recognized in poisonous metals such as mercury — to continually build up in the tissues of higher organisms as they gradually climb the food chain. All of this led to the abandonment of DDT.

There are alternatives to the use of chemical pesticides, such as biological control measures, which function by fighting parasites or weeds with organisms that can eliminate them or halt their reproduction. However, it is necessary to proceed with caution here, because a predator or virus introduced to combat the parasites may lead to unforeseen situations and unexpected results. Foreseeing the future effects of an intervention on nature is not always easy! When the ecological equilibrium is altered, it is difficult to restore it or establish another.

1.12 Major disasters

In order to illustrate how humanity is always and everywhere exposed to serious risks, even of its own device, it is sufficient to evoke the names of some places where major ecological disasters have occurred in recent years. The table below describes some of the major tragedies which have taken place over the past decades. Even disasters of lesser impact can lead to an alteration of the ecological equilibrium.

Table 2: Major disasters

Date	Cause	Location
16 April 1947	The SS Grandcamp, carrying ammonium nitrate fertilizer, exploded in the Texas City harbour, followed the next morning by the explosion of the SS High Flyer. The disaster killed almost 576 and injured several thousand. The explosion was felt 75 miles away in Port Arthur, and created a 15-foot tidal wave.	Texas, USA
1956	Minamata disease officially recognized as mercury poisoning. It developed in people who ate contaminated seafood taken from Minamata Bay and adjacent coastal waters in the period after World War II when methyl mercury was dumped into the sea as an unwanted by-product of acetaldehyde processing at the Chisso industrial plant in Minamata, Japan.	Minamata, Japan
7 October 1957	Fire in the plutonium production reactor. Dispersal of radioactive material into the countryside. 39 deaths due to cancer.	Windscale, north of Liverpool, Great Britain
18 March 1967	The Liberian oil tanker Torrey Canyon sinks and spills 123,000 tons of crude oil. 180 km of English and French beaches are polluted.	Off the shores of Cornwall, Great Britain
10 July 1976	Toxic cloud of dioxin.	Seveso, Italy

Date	Cause	Location
16 March 1978	220,000 tons of crude oil spilled after Amoco Cadiz ran aground near Portsall; the slick eventually covered 125 miles of Breton coast.	Brittany, France
28 March 1979	The most severe nuclear disaster in a civilian plant.	Three-Mile Island, Pennsylvania, USA
3 June 1979	Itox factories explode. More than 600,000 tons of crude oil dispersed.	Gulf of Mexico
19 July 1979	Collision of two Liberian oil tankers, the Atlantic Express and the Aegean Captain. 272,000 tons of petroleum dispersed.	Off the shores of Trinidad and Tobago, Caribbean Sea
7 August 1979	Dispersal of enriched uranium from a radioactive fuel plant. 1,000 people contaminated.	Erwin, Tennessee, USA
2-3 December 1984	The greatest disaster in the history of the chemical industry. Forty tons of a lethal gas mixture discharged from a pesticide production plant.	Bhopal, India
26 April 1986	Meltdown at the nuclear power plant. Formation of radioactive clouds that contaminate various countries.	Chernobyl, Kiev, Ukraine
24 March 1989	The US oil tanker Exxon Valdez runs aground as it attempts to avoid ice blocks. 40,000 tons of crude oil form a black patch covering 4,000 km^2 of water.	25 miles from the Valdez Trans Alaska oil pipeline terminal, Gulf of Alaska.

Date	Cause	Location
4 June 1989	The Ufa train disaster happened near the cities of Ufa and Asha in the Soviet Union on the Trans-Siberian Railway. A liquefied petroleum gas explosion killed 575 people and wounded over 600, making it the most deadly railway accident in Soviet history, as two trains passing each other threw sparks near (1 km) a leaky pipeline. Both trains were carrying children; one returning from a holiday break on the Black Sea, one taking children there. The explosion was so powerful it blew out windows in Asha, eight miles from the epicentre.	Near the cities of Ufa and Asha in the Soviet Union.
9 June 1990	Explosions cause fires on the Norwegian oil tanker Mega Borg. 100,000 tons of crude oil spilled.	Galveston, Texas, USA
10 April 1991	The ferry Moby Prince rams the motor ship Agip Abruzzo. 25,000 tons of petrol oil released and 140 people killed.	Port of Livorno, Italy
11 April 1991	Fire with explosions on the Cypriot oil tanker Haven. 2 dead, 147,000 tons of petrol oil released, 500 km^2 of tar on the sea floor.	Off the shore of Arenzano, Liguria, Italy
3 December 1992	The Greek oil tanker Aegeum Sea hits the port wharf. 79,000 tons of crude oil spilled into the water, forming a patch 30 km long and 2 km wide.	La Coruña, Spain
30 March 1994	The Panamanian oil tanker Seki, with 268,332 tons of crude oil, collides with the Baynunah of the United Arab Emirates, and loses part of its load.	Off the shore of the Port of Fujairah, Persian Gulf

Date	Cause	Location
3 June 1996	Exploding fireworks cause a fire in which thousands of hectares of forest are destroyed.	North of Anchorage, Alaska
September 1997	Fires started by farmers devour hundreds of hectares of forests and generate a toxic cloud extending from Singapore to Indonesia, from Malaysia to the Philippines. Monsoons and dry spells aggravate the situation.	Asia
17 October 1998	Pipeline at Jesse Nigeria exploded, instantly killing more than 500 people and severely burning hundreds more. The final death toll was 1200 people. Up to 2000 people had been lining up with buckets and bottles to scoop up oil. The fire spread and engulfed the nearby villages of Moosqar and Oghara, killing farmers and villagers sleeping in their homes.	Niger Delta, Nigeria
30 September 1999	Because of a multiplication error, a uranium leak in a nuclear reaction in a private laboratory contaminates about 50 people.	Tokaimura, Japan
21 September 2001	Explosion at Azote de France (AZF) agricultural chemicals factory near Toulouse. 31 people dead, at least 650 people hospitalized.	Toulouse, France
11 May 2004	ICL Plastics plc Stockline Plastics plant in Glasgow explodes, killing 9 and injuring more than 40.	Glasgow, Scotland
20 October 2004	Gas explosion in Daping coal mine in Henan province killed 148 people and injured many others.	Daping coal mine in Henan province, China

1.13 Species extinction

Many biotopes of the animal and plant kingdoms are jeopardized in their natural form by biotope alterations, soil exhaustion and dangerous substance emissions. Because of the extinction of countless animal and plant species due to worsening environmental conditions, the biosphere gradually loses a vast part of its genetic variety, which has developed over millions of years. This genetic variety, however, is irreplaceable for the support systems of life on the earth, since it makes evolutionary adaptations within the biosphere possible. It is equally irreplaceable for medical research, zootechnics and crop cultivation. Almost half of all medical drugs being sold have their origin in chemical substances which were first obtained from nature.

The destruction and pollution of the tropical forests and other natural environments leads to the extinction of various animal and plant species. The risk is no longer for single, symbol species such as pandas in China, white oryxes in Arabia, snow leopards in Asia, monk seals in the Mediterranean, or bald eagles in the United States, but extinction appears close at hand for animals whose 'hordes' once seemed 'endless,' such as zebras in Africa and bison in North America. Another problem has been developing over the past 40 years with intensive fishing. This, together with water pollution, has caused alarming drops in otherwise immense populations of anchovies, sardines and codfish. Furthermore, along the South American Pacific coasts, excessive fishing has corresponded to a decrease in marine birds, whose food is now depleted, with consequent damage to the market for their guano. The result is that the exploitation of one species leads to the extinction or reduction of other species as well.

1.14 Noise pollution

Noise pollution refers to the introduction of noise into the domestic or external environment such that it bothers or disturbs rest and other human activities, constitutes a danger to human health, contributes to the degradation of ecosystems, material goods, monuments, the home or external environment, or such that it interferes with the legitimate enjoyment of the environments themselves. The most common sources noise worldwide are transportation systems, motor vehicle noise, but also aircraft noise and rail noise. Poor urban planning may give rise to noise pollution, since side–by–side industrial and residential buildings can result in noise pollution in the residential area. Other sources are office equipment, factory machinery, construction work, appliances, power tools, lighting hum and audio entertainment systems.

Noise can affect are both health and behaviour. The unwanted sound can damage physiological and psychological health. Noise pollution can cause annoyance and aggression, hypertension, high stress levels, tinnitus, hearing loss, and other harmful effects. Furthermore, stress and hypertension are the leading causes of health problems, whereas tinnitus can lead to forgetfulness, severe depression and at times panic attacks. The mechanism for chronic exposure to noise leading to hearing loss is well established. The elevated sound levels cause trauma to the cochlear structure in the inner ear, which gives rise to irreversible hearing loss. A very loud sound in a particular frequency range can damage the cochlea's hair cells that respond to that range thereby reducing the ear's ability to hear those frequencies in the future. However, loud noise in any frequency range has deleterious effects across the entire range of human hearing.

The meaning listeners attribute to the sound influences annoyance, so that, if listeners dislike the noise content, they are annoyed. What is music to one is noise to another. If the sound causes activity interference (for example, sleep disturbance), noise is more likely to annoy. If listeners feel they can control the noise source, the noise is less likely to be annoying. If listeners believe that the noise is subject to third–party control, including police, but control has failed, they are more annoyed. The inherent unpleasantness of the sound causes annoyance. If the sound is appropriate for the activity it is in context. If one is at a race track the noise is in context and the psychological effects are absent. If one is at an outdoor picnic the race track noise will produce adverse psychological and physical effects.

Technology to mitigate or remove noise can be applied in various ways. There are a variety of strategies for mitigating roadway noise including: use of noise barriers, limitation of vehicle speeds, alteration of roadway surface texture, limitation of heavy duty vehicles, use of traffic controls that smooth vehicle flow to reduce braking and acceleration and tyre design. An important factor in applying these strategies is a computer model for roadway noise; this system is capable of addressing local topography, meteorology, traffic operations and hypothetical mitigation. Costs of building–in mitigation can be modest, provided these solutions are sought in the planning stage of a roadway project. Aircraft noise can be reduced to some extent by design of quieter jet engines, which was pursued vigorously in the 1970's and 1980's. This strategy has brought limited but noticeable reduction of urban sound levels. Reconsideration of operations, such as altering flight paths and time of day runway use, have demon-strated benefits for residential populations near airports. Exposure of workers to industrial noise has been addressed since the 1930's. Changes include redesign of industrial equipment, shock mounting assemblies and physical barriers in the workplace. Solutions can include acoustic isolation of buildings in noisy office zones.

1.15 The myth of overpopulation

The world population consisted of about 6 billion people in the year 2000. The problem concerns the distribution of the population, not their number. The argument by which these numbers are used to encourage and justify the use of artificial birth control is entirely wrong. Currently, some scientists are beginning to refute the myth of the need for so–called population control.[31] Population control, in fact, is based on an economic illusion. With fewer clients, public services (trains, buses) are faced with diseconomies. Consequently, for example, some hospitals and schools would have to close due to a lack of personnel. Further-more, where the population is controlled (in many parts of Europe), the problem has been emerging of a continually aging population. In reality, the population increase creates a less difficult life, as an impartial analysis of economic and political history demonstrates. The signs of a growing economy generally coincide with population growth; this fact has been formalized in several economic models, such as that of the Polish economist Kalecki.

A significant incoherence on the part of certain environ-mentalists becomes apparent, for while they are against undue manipulation of the micro–environment of the universe, they not only do not oppose but in fact support unjust manipulation of the human body.[32] This incoherence becomes obvious, for example, when so–called 'green' groups favour birth control through artificial contraception:

> To pollute the waterways of the human body with chemicals, and block its passages with metal and plastic barriers, deliberately to prevent its functioning in a normal and healthy way, is an extension of the industrial mentality to the most private human sphere. It places woman in the hands of technocrats and the big corporations, totally dependent on them for as long as she wishes to remain the sexual plaything of the men who refuse to take responsibility for her children.[33]

Thus, it is entirely unacceptable to follow the approach of certain ecologists who see the solution to the crisis of humanity and its environment in birth control. These ideological manipulations have an egoistic philosophy at their roots which, in reality, seeks to make life ever more pleasurable for rich countries while ignoring underdeveloped areas. In some of the poorest countries, birth control often deprives cities of the manpower necessary for the development and care of the environment. The systematic family planning campaign is a new form of oppression: 'It is the poorest populations which suffer such mistreatment, and this sometimes leads to a tendency towards a form of racism, or the promotion of certain equally racist forms of eugenics.'[34] Nonetheless, the problem of overpopulation is flaunted by experts who are often motivated by ideological considerations. A recent editorial in an important newspaper expressed a healthy scepticism about the ideological speculations concerning overpopulation:

> Monaco is one of the most populated territories on

Earth, with a population of 40,112 to the square mile, followed closely by Hong Kong. Neither is notably less successful or unhappier than the United Kingdom, which has a population density between 25 and 70 times less than theirs. The most sparsely populated territories of the world, from Antarctica to the wilder Highlands of Scotland, are not notable for their quality of life, opportunities for the young or the number of volunteers to live there permanently.[35]

The Holy See is particularly concerned by strategies which target population reduction as the primary means for overcoming ecological problems. Programs for population reduction, directed and financed by the developed Northern nations, become an easy substitution for justice and development in developing Southern countries. These programs evade the questions of distributive justice and the development of the Earth's abundant resources. In many occasions, the Holy See has expressed its opposition to the establishment of quantitative population aims and objectives, which constitute a violation of human dignity and rights.[36]

The real problem of the poorest countries is not so–called overpopulation, but the egoism of the West. Population reductions in the poorest countries, in fact, would lead them to even more acute poverty. A variety of people also leads to creativity within a society. Instead, many environmentalists give the impression that they believe human beings are a 'scar' or 'cancer' of the planet, a violation of the otherwise perfect natural order. This idea finds no support in revelation, where the truth is instead discovered to be the exact opposite: man was placed on Earth by God, and was ordered to 'be fertile and multiply, to fill the

earth and subdue it' (Genesis 1:28). Considering the existence of other people a misfortune or even a violation of
nature is an idea that radically distances itself from Judeo–
Christian ethics. We were created in the image and likeness
of God, and this means that every human being in the
world is sacred as such, since he or she adds an inestimable
value to creation which did not exist before. The idea that
people are simply a drain on energy resources not only
contradicts our faith, but negates even the true contribution of human beings to the common good of human
society and to the entire environmental realm. Any understanding which does not welcome a new human person
with joy, based on that person's intrinsic value and what
that person could contribute to the world, from a providential point of view, is in fundamental contradiction with
Catholic ethics.

Notes

1   Concerning this idea, see the article in *Corriere della Sera*
     114/23 (17 January 1989), p. 17.

2   This is abbreviated SAR, and is measured in Watts per
     kilogram (W/kg).

3   See J. M. Delgado, J. Leal, J. L. Monteagudo, M. G. Gracia,
     'Embryological changes induced by weak, extremely low
     frequency electromagnetic fields' in *Journal of Anatomy*
     134/3 (May 1982), pp. 533–551. See also J. D. Harland, R.
     Liburdy, 'Environmental magnetic fields inhibit the antipro-
     liferative action of tamoxifen and melatonin in a human
     breast cancer cell line' in *Bioelectromagnetics* 18/8 (1997), pp.
     555–562.

4   See S. Aalto, C. Haarala, A. Bruck, H. Sipila, H. Hamalainen,
     J. O. Rinne, 'Mobile phone affects cerebral blood flow in

humans' in *Journal of Cerebral Blood Flow & Metabolism* 26/7 (July 2006), pp. 885–890; M. Koivisto, C. M. Krause, A. Revonsuo, M. Laine, H. Hamalainen, 'The effects of electromagnetic field emitted by GSM phones on working memory' in *Neuroreport* 11/8 (June 2000), pp. 1641–1643.

5   See V. Binhi, *Magnetobiology: Underlying Physical Problems.* (New York: Academic Press, 2002).

6   N. Wertheimer, E. Leeper, 'Electrical wiring configurations and childhood cancer' in *American Journal of Epidemiology* 109(1979), pp. 273–284.

7   See P. Brodeur, *Currents of Death: Power Lines, Computer Terminals, and the Attempt to Cover Up the Threat to Your Health* (New York: Simon and Schuster, 1989); Idem, *The Great Power Line Cover–Up: How the Utilities and Government Are Trying to Hide the Cancer Hazard Posed by Electromagnetic Fields* (Little, Brown and Company: 1993, hardback). There is also a 1995 paperback edition.

8   A. Ahlbom, E. Cardis, A. Green, M. Linet, D. Savitz, A. Swerdlow, 'Review of the Epidemiologic Literature on EMF and Health' in *Environmental Health Perspectives* 109/S6 (December 2001).

9   G. Draper, T. Vincent, M. E. Kroll, J. Swanson, 'Childhood cancer in relation to distance from high voltage power lines in England and Wales: a case–control study,' in *British Medical Journal* 330 (2005), pp. 1290f.

10   See P. Fews, D. Henshaw, P. Keitch, J. Close, R. Wilding, 'Increased exposure to pollutant aerosols under high voltage power lines' in *International Journal of Radiation Biology* 75/12 (December 1999), pp. 1505–1521; P. Fews, D. Henshaw, R. Wilding, P. Keitch, 'Corona ions from power-lines and increased exposure to pollutant aerosols' in *International Journal of Radiation Biology* 75/12 (December 1999), pp. 1523–1531.

11   World Health Organization, *Electromagnetic fields and public health*, Fact sheet N° 322 (June 2007).

12 Health Protection Agency, *Power Frequency Electromagnetic Fields, Melatonin and the Risk of Breast Cancer* (February 2006).

13 See Independent Expert Group on Mobile Phones, *The Stewart Report* (Didcot 2000), 5.266–5.268: 'There is also good evidence that exposure to mobile phone signals …has direct, short–term effects on the electrical activity of the human brain and on cognitive function. These could have their origin in a variety of biological phenomena, for which there is some evidence from experiments on isolated cells and animals. There is an urgent need to establish whether these direct effects on the brain have consequences for health, because, if so, and if a threshold can be defined, exposure guidelines will have to be reconsidered. It is also important to determine whether these effects are caused by local elevation of temperature or, as seems possible, by some other, 'non–thermal', mechanism. The epidemiological evidence currently available does not suggest that Radio Frequency exposure causes cancer. This conclusion is compatible with the balance of biological evidence, which suggests that Radio Frequency fields below guidelines do not cause mutation, or initiate or promote tumour formation. However, mobile phones have not been in use for long enough to allow a comprehensive, epidemiological assessment of their impact on health, and we cannot, at this stage, exclude the possibility of some association between mobile phone technology and cancer. In view of widespread concern about this issue, continued research is essential. Experimental studies on cells and animals do not suggest that mobile phone emissions below guidelines have damaging effects on the heart, on blood, on the immune system or on reproduction and development. Moreover, even prolonged exposure does not appear to affect longevity. The limited epidemiological evidence currently available also gives no cause for concern about these questions.'

14 Defence Intelligence Agency, *Biological effects of electromagnetic radiation (radiowaves and microwaves) – Eurasian*

Communist Countries. DST–1810S–074–76, March 1976.

[15] See Final Report, *Risk Evaluation of Potential Environmental Hazards from Low Frequency Electromagnetic Field Exposure using Sensitive in vitro Methods* (European Union, 2004).

[16] See A. Bouville and W. M. Lowder, 'Human Population Exposure to Cosmic Radiation' in *Radiation Protection Dosimetry* 24/1 (1988), pp. 293–299.

[17] See T. Larssen, et al., 'Acid Rain in China' in *Environmental Science and Technology* 40/2 (2006), pp. 418–425.

[18] Some data can provide an idea on the biodegradation times of various waste products, or rather the amount of time nature requires to get rid of them: a paper tissue: 4 weeks; a daily newspaper: 6 weeks; cloth and wool: 8–10 months; a magazine with glossy pages: 8–10 months; a match: 6 months; a cigarette butt: 1 year or more; chewing gum: 5 years; an aluminium can: 10 years; a classic (non–biodegradable) plastic bag: 500 years or more; synthetic fabric: 500 years or more; a plastic bottle: almost 100 years; lighters: 100 years; tampons and diapers: 200 years; phone cards: 1000 years; glass bottles: undetermined.

[19] This information can seem to be alarmism, if it is not recalled that ¾ of the Earth's surface, or 75%, is covered by ocean. Thus, the fact that forests are 'only' 7% of the remaining 25%, which is more than ¼, appears somewhat less alarming.

[20] See W. Rodhe, 'Crystallization of eutrophication concepts in North Europe' in *Eutrophication, Causes, Consequences, Correctives* (Washington D.C.: National Academy of Sciences, 1969), pp. 50–64.

[21] B. J. Soden and I. M. Held, 'An Assessment of Climate Feedbacks in Coupled Ocean–Atmosphere Models' in *Journal of Climate* 19 (2005).

[22] Intergovernmental Panel on Climate Change, *Climate Change 2007: The Physical Science Basis. Contribution of Working Group I to the Fourth Assessment Report of the Intergovernmental Panel on Climate Change*, 'Summary for

Policymakers', p. 5.

23   *Ibid.*, p. 10.

24   G. C. Hegerl, et al. 'Understanding and Attributing Climate Change' in Intergovernmental Panel on Climate Change, *Climate Change 2007: The Physical Science Basis. Contribution of Working Group I to the Fourth Assessment Report*, pp. 681–682. See also C. M. Ammann et al., 'Solar influence on climate during the past millennium: Results from transient simulations with the NCAR Climate Simulation Model' in *Proceedings of the National Academy of Sciences of the United States of America* 104/10 (2007), pp. 3713–3718; S. K. Solanki, I. G. Usoskin, B. Kromer, M. Schüssler & J. Beer, 'Unusual activity of the Sun during recent decades compared to the previous 11,000 years' in *Nature* 431 (28 October 2004), pp. 1084–1087.

25   See Intergovernmental Panel on Climate Change, *Climate Change 2007: The Physical Science Basis*, 'Summary for Policymakers', pp. 5–6.

26   See D. H. Douglass, B. D. Pearson, S. F. Singer, 'Altitude dependence of atmospheric temperature trends: Climate models versus observation' in *Geophysical Research Letters* 31/ L13208 (9 July 2004).

27   See D. H. Douglass, B. D. Pearson, S. F. Singer, P. C. Knappenberger, P. J. Michaels, 'Disparity of tropospheric and surface temperature trends' in *Geophysical Research Letters* 31/ L13207 (9 July 2004).

28   J. R. Christy, 'My Nobel Moment' in *Wall Street Journal* (1 November 2007), p. A19.

29   Mgr. Pietro Parolin, *Address at the 62nd Session of the General Assembly of the United Nations* during the high–level event on climate change entitled 'The future is in our hands: addressing the leadership challenge of climate change' (24 September 2007).

30   See N. Lawson, *An Appeal to Reason: A Cool Look at Global Warming* (London: Duckworth, 2008). Lawson proposes a

rational response to global warming, and explains why the mistaken conventional wisdom has become the quasi-religion it is today, and the dangers that this presents.

31 See for example: J. L. Simon, *The State of Humanity* (Oxford: Blackwell, 1995) and Idem, *The Ultimate Resource* (Princeton: Princeton University Press, 1996). These two books show that population growth is not a negative factor from the economic point of view. The neo–Malthusian approach as a solution to neo–eschatological hypotheses connected to the population bomb was initially propagandized by the so-called 'Club of Rome,' at the end of the 1960's, through the works *Limits to Growth* and *Toward Global Equilibrium* which were developed with great publicity 'campaigning' by the M.I.T. System Dynamic Group and *Strategy for Survival* by M. Mesarovic and E. Pestel; see also W. Leontief, A. P. Carter and P. A. Petri in *The Future of World Economy* (United Nations: 1977). A founded and incisive criticism of these theses, which time has corroborated, is found in the text by C. Clark, *The myth of over–population* (Melbourne: Advocate Press Pty. Ltd., 1973), also available in Italian (Milan: Ares, 1974); another important text, though 'dated', is by J. Verrière, *Les politiques de population* (Paris: Presses Universitaires de France, 1978). In effect, in relation to the myth of the exhaustion of raw materials and energy in the near future, in conjunction with the imminent (and 'inevitable') environmental crisis, one cannot help but recall — against all of the ideological rubbish in circulation — the pondered and documented observations of H. Jonas in *The Imperative of Responsibility: In Search of an Ethics for the Technological Age* (Chicago: The University of Chicago Press, 1984), according to which the only serious world problem (which must still be verified and studied) might be a non-eliminable excess of heat, deriving from the degradation of the energy that we utilize, in accordance with the principle of entropy.

32 On this topic, see chapter two below. See also S. L. Jaki,

'Ecology or Ecologism?' in G. B. Marini–Bettòlo (ed.), *Man and his Environment. Tropical Forests and the Conservation of Species* (Vatican City State: Pontifical Academy of Sciences, 1994), pp. 271–293. Some scholars, such as Anna Bramwell, make a distinction between environmentalists and 'ecologists' (based on a distinction between environmentalism and ecologism), sustaining that 'ecologists' are more extreme in demanding a total elimination and replacement of the current political system to accommodate their aims, predicting total catastrophe if everyone does not join them. Environmentalists, in this dichotomy, are more moderate, looking for ways to solve particular environmental problems without such sweeping forecasts of Armageddon. See also R. P. McIntosh, 'History of Ecologism?' in *Ecology* 70/6 (December 1989), pp. 1963–1964. For the purposes of this book, however, the terms 'environmentalism' and 'ecologism,' and their respective derivatives 'environmentalist' and 'ecologist,' are used interchangeably to refer to an ideological approach to environmental issues which ignores the centrality of human beings, subordinating them to the environment in one way or another.

33   S. Caldecott, 'Cosmology, eschatology, ecology: some reflections on *Sollicitudo Rei Socialis*,' in *Communio* 15/3 (1988), p. 313.

34   Pope John Paul II, Encyclical Letter *Sollicitudo Rei Socialis*, 25.

35   Editorial in: *The Times* (10 August 1993), p. 15.

36   R. Martino, The Holy See at the Rio Conference, *Stewardship and Solidarity* – Summary document (5 June 1992), 9.

2

Ecology or Ideology?

No one can put together what has crumbled into dust, but
You can heal men whose conscience has become twisted; You
give the soul its former beauty, which long ago it had lost
without a hope of change. With You, nothing is hopeless. You
are Love. You are the Creator and the Redeemer of all things.
We praise You with this song: Alleluia!

Metropolitan Tryphon, Creation Akathistos, 10

2.1 The phenomenon and its interpretation

The relationship between cause and effect is not always
clear and immediate in environmental issues when dealing, for example, with the greenhouse effect or the hole in
the ozone layer.[1] On the one hand, economic interests come
into play; on the other, the fanaticism of environmentalists.
Who is truly concerned with environmental problems?
Certainly the Catholic Church is, as we will see later on in
chapter three. The debate concerning the quality of life
cannot reduce this quality to a merely natural and physical
level, as do many political groups such as the Green,
Socialist, and Communist parties in order to further their

own aims. A structural change, in fact, is not always the same as an improvement in living conditions. Ecology has thus been transformed from a particular science into a general science of bio–cultural existence, and thence into an ideology which we can call ecologism.[2] Stanley Jaki points out how the shift 'from physics to physicalism and from science to scientism may provide an informative parallel with the shift from ecology to ecologism'.[3] Ecologies that seemingly begin with the program of saving man's environment quickly run their logic to the point where the environment takes absolute priority over man. This ideology easily takes root in Darwinist circles where man is seen to be the product of purely natural forces.[4] Part and parcel of this pernicious view is the erroneous claim that man is simply one of a very large number of species, all equally valuable and enjoying the same rights.[5]

Political ecological ideology comes in various shapes and forms. First *ecocapitalism* exists despite the fact that most conservatives believe that environmentalism is a lot of hype, and that scientific advance will fix all ecological problems. However libertarian environmentalists propose a *Coasian* solution to the problem.[6] That is, they believe that the problem with the environment is that it has not been divided up into property. Supposedly, if we sold all of the air, water, and land to private concerns, then rights to pollute could be bought and sold, perfectly balancing industrial and environmental interests. Another variant of political ecology is *conservationism*, proposed by the Sierra Club and other pre–60's environmental groups. These were largely made up of hunters and outdoorsmen who were concerned about preserving 'wildlife' and 'the great outdoors'. Conservationists established the national parks, and still get excited about preserving America's vital resources.

Environmentalism is an approach oriented toward mailing–list memberships, well–paid central staffs, and legal and lobbying activities, instead of grass–roots activism. These lobbies include eco–capitalists and conservationists, and often accept corporate contributions and board members. While most of their members and activities are implicitly anti–corporate, they are loath to openly proclaim a radical stance. *Ecopopulism* represents a popular approach to the environment. Its supporters are the mothers enraged about their children's illnesses who organize a toxic waste protest, or the workers who get the shop steward to contact the Occupational Safety and Health Administration. Except for the unions, grassroots groups have weak central staffs if any. While these groups are non–ideological they express strongly anti–corporate views, since they are directly confronted with the incompatibility of profiteering and human needs. Their militancy is often undercut, however, by their communities' dependence on the jobs and tax–base of the companies they are fighting. They also usually lack a broader analysis of ecological politics, and fall into the parochialism of *Not In My Back Yard*.

The expression *Greens* came into vogue with the ascension of the West German Green Party into the *Bundestag,* and the subsequent emergence of Green parties throughout Europe and also in developing countries. People who call themselves Greens generally advocate multi–issue, independent political action through green parties, but range from those strictly concerned with an ecological agenda, to revolutionary anarchist greens who see electoral politics as only propaganda, to 'red greens' who believe green politics has replaced Marxism as a comprehensive radical ideology, unifying socialism, feminism, anti–racism, and so forth.

Deep ecology is an anti–rationalist philosophy, and therefore difficult to define. However, its principal tenet is the replacing of anthropocentric thinking with biocentrism. Biocentrism views 'Nature' as valuable in itself, and that all species are equally valuable within it. Deep ecologists argue for a radical reduction in human population, in human interference in nature, and in the human standard of living. They tend to argue that pre–industrial peoples are in an organic harmony with the natural order, and that European industrial culture has severed this harmony. For this ideology, industrial society is like a cancer spreading through a global host. Deep ecologists overlap with the New Age Greens, who are more concerned with lifestyle changes, self–realization and spirituality than political change. Nevertheless there are anarchistic, eco–guerrilla exponents of deep ecology who have blown up construction equipment and spiked trees to stop logging.[7] *EcoMarxists* continue the Marxist tradition of arguing that all social problems result from capitalism. They often argue somewhat naively that the miserable ecological disasters of the Communist regimes resulted from their adopting 'capitalist technology'. *Ecofeminists* tend to be New Age and deep ecologist, though more radical, weaving their insights into a non–linear critique of the entire patriarchal, logocentric, European worldview. Their basic point is that patriarchal society associates women with nature, and rapes and debases both. The ecological movement, therefore, must include the overturning of patriarchy. Ecofeminism grew out of women's anti–

military mobilizations and has strong roots in the radical witchcraft movement.

Bioregionalists believe that human societies should be decentralized, and political boundaries should reflect bio–geographic locales. Instead of America and Canada with states and provinces, we should have the Great Lakes BioRegion, and then smaller bioregions around water-sheds, valleys and so on. Bioregionalists tend to be subculturally New Age and deep ecological, and uninter-ested in political activism.

Social Ecology purports to be a coherent philosophy of ecological anarchism. Here, humanity is not seen as sepa-rate from nature, but rather 'nature aware of itself'. Un–natural hierarchy has arisen in society, however, and caused alienation between human beings, and between humanity and nature, cutting us off from our oneness. The way to save the ecosystem, therefore, is to dismantle human hierarchy in all its forms, including race, sex, and class, which will return us to a natural ecological sanity. The only form of State that social ecologists abide is the city–state, where small size allows all decisions to be made through Grecian direct participation. Social ecologists are anti–capitalist, and advocate the municipalization of the economy.

Ecosocialists propose that democratic socialism is a necessary condition for ecological protection, though not a sufficient one. Ecosocialists point to the ecological successes of democratic socialist governments of Northern Europe, where workers' parties and unions were powerful enough to establish policies opposed by corporations. Ecosocialists contrast these socialist successes to the disas-ters of Communism, which completely forbade opposition to the bureaucrats' industrial plans, and to the marginal

successes of democratic capitalism, which allowed democratic opposition to the industrial system, but limited the
permissible interference with the prerogatives of capital.
Ecosocialists reject the idea that socialism by itself will save
the ecosystem, and believe that only a broad coalition of
the 'democratic left', including ecological groups and other
movements, can establish a just and sustainable society.

Ecofascism has a well–established tradition, going back
well beyond the time of Hitler. The ecological components
of Nazism, their central role in Nazi ideology and their
practical implementation during the Third Reich are well
known. Germany is not only the birthplace of the science
of ecology and the homeland of Green politics' rise to
prominence; it has also been home to a peculiar synthesis
of naturalism and nationalism forged under the influence
of the Romantic tradition's anti–Enlightenment irrational-
ism. Two nineteenth century figures exemplify this
ominous conjunction: Ernst Moritz Arndt and Wilhelm
Heinrich Riehl.

While best known in Germany for his fanatical nationalism, Arndt was also dedicated to the cause of the
peasantry, which lead him to a concern for the welfare of
the land itself. Historians of German environmentalism
mention him as the earliest example of 'ecological'
thinking in the modern sense.[8] Arndt's environmentalism,
however, was inextricably bound up with virulently xeno-
phobic nationalism. At the very outset of the nineteenth
century the deadly connection between love of land and
militant racist nationalism was firmly set in place. Riehl, a
student of Arndt, further developed this sinister tradition.
In some respects his 'green' streak went significantly
deeper than Arndt's; presaging certain tendencies in recent
environmental activism. Even here nationalist pathos set

the tone: 'We must save the forest, not only so that our ovens do not become cold in winter, but also so that the pulse of life of the people continues to beat warm and joyfully, so that Germany remains German.'[9] Riehl was an implacable opponent of the rise of industrialism and urbanization; his overtly antisemitic glorification of rural peasant values and undifferentiated condemnation of modernity established him as the 'founder of agrarian romanticism and anti–urbanism.'[10]

The emergence of modern ecology forged the final link in the fateful chain which bound together aggressive nationalism, mystically charged racism, and environmentalist predilections. The German word *Oekologie* appeared in 1860; some say that a certain W. Reiter coined it. The term was used by the German biologist E. Haeckel in 1866 to indicate the study of an organism's relation to the exterior surrounding world, that is, in a broad sense, the study of the conditions of existence.[11] He developed what was later termed *Haeckel's law of recapitulation* according to the principle that 'ontogeny recapitulates phylogeny', and was first to draw up a genealogical tree relating the various orders of animals. As a philosopher he was an exponent of monistic philosophy, which postulated a totally materialistic view of life as unity and which he presented as a necessary consequence of the theory of evolution.

Haeckel's contributions to zoological science were a mixture of sound research and speculations often with insufficient evidence (including use of forged drawings). His law is now discredited and some of his theses became a part of the pseudoscientific basis for Nazism. To observe living beings and their environment, ecology explores every aspect of nature, and in so doing it makes use of all the other sciences. Haeckel was also the chief popularizer of Darwin and evolutionary theory for the German–speaking world. He affirmed nordic racial superiority, strenuously opposed race mixing and enthusiastically supported racial eugenics. His fervent nationalism became fanatical with the onset of World War I, and he fulminated in antisemitic tones against the post–war Council Republic in Bavaria. In this way Haeckel 'contributed to that special variety of German thought which served as the seed bed for National Socialism. He became one of Germany's major ideologists for racism, nationalism and imperialism.'[12] Near the end of his life he joined the Thule Society, 'a secret, radically right–wing organization which played a key role in the establishment of the Nazi movement.'[13]

The philosopher Ludwig Klages profoundly influenced the German youth movement and particularly shaped their ecological consciousness. He authored an important essay titled 'Man and Earth' which anticipated most of the themes of the contemporary ecology movement.[14] It decried the accelerating extinction of species, disturbance of global ecosystemic balance, deforestation, destruction of aboriginal peoples and of wild habitats, urban sprawl, and the increasing alienation of people from nature. In emphatic terms it disparaged Christianity, capitalism and the ideology of 'progress'. It even condemned the environmental destructiveness of rampant tourism and the slaughter of whales, and displayed a clear recognition of the planet as an ecological totality.

Another philosopher who helped bridge fascism and environmentalism was Martin Heidegger. A much more renowned thinker than Klages, Heidegger preached 'authentic Being' and harshly criticized modern technology, and is therefore often celebrated as a precursor of ecological thinking. On the basis of his critique of technology and rejection of humanism, contemporary deep ecologists have elevated Heidegger to their pantheon of eco–heroes. Heidegger's critique of anthropocentric humanism, his call for humanity to learn to let things be, his notion that humanity is involved in a 'play' or 'dance' with earth, sky, and gods, his meditation on the possibility of an authentic mode of 'dwelling' on the earth, his complaint that industrial technology is laying waste to the earth, his emphasis on the importance of local place and 'homeland', his claim that humanity should guard and preserve things, instead of dominating them — all these aspects of Heidegger's thought help to support the claim that he is a major deep ecological theorist.[15] As for the philosopher of Being himself, he was — unlike Klages, who lived in Switzerland after 1915 — an active member of the Nazi party and for a time enthusiastically, even adoringly supported the Führer. His mystical panegyrics to *Heimat* (homeland) were complemented by a deep antisemitism, and his metaphysically phrased broadsides against technology and modernity converged neatly with populist demagogy. Ernst Lehmann was a professor of botany who characterized National Socialism as 'politically applied biology':

> We recognize that separating humanity from nature, from the whole of life, leads to humankind's own destruction and to the death of nations. Only through a re–integration of humanity into the whole of nature can our people be made stronger. That is the fundamental point of the biological tasks of our age. Humankind alone is no longer the focus of thought, but rather life as a whole... This striving toward connectedness with the totality of life, with nature itself, a nature into which we are born, this is

the deepest meaning and the true essence of National Socialist thought.[16]

This unmediated adaptation of biological concepts to social phenomena served to justify not only the totalitarian social order of the Third Reich but also the expansionist politics of *Lebensraum* (the plan of conquering 'living space' in Eastern Europe for the German people). It also provided the link between environmental purity and racial purity.

No aspect of the Nazi project can be properly understood without examining its implication in the Holocaust. Here, too, ecological arguments played a crucially malevolent role. Not only did the 'green wing' refurbish the sanguine antisemitism of traditional reactionary ecology; it catalysed a whole new outburst of lurid racist fantasies of organic inviolability and political revenge. The confluence of anti–humanist dogma with a fetishization of natural purity provided not merely a rationale but an incentive for the Third Reich's most heinous crimes.

Even certain pseudo–mystical sects take an interest in ecology, such as the so–called New Age or Next Age movement, which is an old enemy in a new form. Some non–Christian religions, such as Buddhism, expound environmental ideas, but many times with a pantheistic approach. It is no accident, therefore, that for many ideologists the first law of ecology is that 'everything is connected to everything else'.[17] The New Age movement envisions unity among all peoples, founded not on God but on merely human values, or, at worst, on evil ideas. Already Haeckel was himself somewhat allied with one of the main exponents of theosophy, Rudolf Steiner.[18] Steiner linked Haeckel with his theosophical ideology:

Theosophical cosmology is a self–contained whole,

derived from the wisdom of the most developed seers. If I had a little more time I would be able to indicate to you how certain natural scientific facts are conducive to testifying to the accuracy of this image of the world. Look at Haeckel's famous phylogenic trees, for example, in which evolution is materialistically explained. If instead of matter you consider the spiritual stages, as Theosophy describes them, then you can make the phylogenic trees as Haeckel did — only the explanation is different.[19]

Often, ecologism is connected to ideologies which conflict with the Catholic faith. It is therefore necessary to be careful: ecology is a science, but ecologism is a mere ideology.

Even some Catholics have recently fallen into ideology by promoting Creation spirituality, or a 'new cosmology' of the Universe, parts of which are variously called the 'new story', 'Earth story', or 'Universe story'. Generally speaking, today's Creation spirituality movement seeks to integrate elements of pagan religions and the traditions of global indigenous cultures with the emerging scientific understanding of the Universe. In the interests of promoting a more Earth–centred view of life, some expressions of Creation spirituality have by–passed out the need for the Redemption of man and woman through Jesus Christ, and have clouded the understanding of the relation between God, humankind, and the world.

The French Jesuit Pierre Teilhard de Chardin (1881–1955) is one of the most well–known theologians influential in ecological ideology. However, his writings were condemned by the Holy See. He attempted to create a fusion of Christianity and evolutionary theory, but taught not so much Catholicism as New Age pantheism. His error starts with a confusion between matter and spirit, whereby even material entities are endowed with spiritual properties: 'We are logically forced to assume the existence in rudimentary form of some sort of psyche in every corpuscle, even in those whose complexity is of such a low or modest order as to render it imperceptible.'[20] This error of panpsychism is followed by a confusion between God and His creation, leading to pantheism. Teilhard described his view of reality as a 'superior form of pantheism' or as an 'absolutely legitimate pantheism'.[21] He admitted to being 'essentially pantheist', and as having dedicated his life to promoting a true 'pantheism of union'.[22] Teilhard goes even further when he denies the immutability of God: 'As a direct consequence of the unitive process by which God is revealed to us, he in some way 'transforms himself' as he incorporates us... I see in the World a mysterious product of completion and fulfillment for the Absolute Being himself.'[23] The concept of Creation is no longer applied in a biblical sense, and Teilhard explicitly stated: 'I find myself completely unsympathetic to the Creationism of the Bible... I find the Biblical idea of creation rather anthropomorphic.'[24] Teilhard stresses instead the mutual complementarity of the Creator and His creation: 'Truly it is not the notion of the contingency of the created, but the sense of the mutual completion of God and the world that makes Christianity live.'[25] God's freedom to create is not clear enough; the cosmos seems to be necessary rather than contingent. Moreover, man's freedom is not clear.

Teilhard proposed a new cosmic Christology in which Christ remains too immanent and does not transcend the evolutionary process; furthermore, Teilhard does not take original sin and the Cross into sufficient consideration. In this understanding, man's cooperation is missing; that is, man is not seen as bringing redemption to all of creation. The Incarnation and Redemption are thus reduced to the natural order, and become necessary rather than gratuitous: 'God cannot appear as the Prime Mover toward the future without becoming Incarnate and without redeeming, that is without Christifying Himself for us.'[26] The Incarnation seems to be a fruit of the evolutionary process: 'Christ is the end–product of the evolution, even of the natural evolution of all beings; and therefore evolution is holy.'[27] Teilhard's conception of evil as a failing and not a condition leads to serious problems with his approach to original sin.[28] This leads to a false idea concerning the Cross and the Redemption wrought by Christ. Briefly, according to Teilhard, the concept of a Cross of expiation is replaced by the notion of a 'cross of evolution' with Christ conceived as the apex of man's spiritual evolution.[29] The angelic world seems to have no place in Teilhard's system. His eschatology is vague to say the least, and tinged with an evolutionary and Hegelian ideology. The term of his continuous creation in Christ is the *Pleroma*, the final state of the world, the consummation of all things in Christ. God's continuous creation is directed to 'the quantitative repletion and the qualitative consummation of all things... the mysterious Pleroma in which the substantial One and the created many fuse without confusion into a whole which, without adding anything essential to God, will nevertheless be a sort of triumph and generalization of Being.'[30] As a result of these errors and ambiguities, the Church has on several occasions drawn attention to the problems and advised vigilance on the part of the faithful.[31]

The leading proponents of the new cosmology and other forms of neo–paganism also include Thomas Berry, Rosemary Radford Ruether and Matthew Fox. While superficially they seem to raise some legitimate concerns, as a rule their teachings run counter to Holy Scripture and to the Tradition of the Church. They mislead many Christians who are seeking the authentic teachings of the Bible and Church Tradition on issues of ecology and environmental justice. For example, dissident Catholic priest Thomas Berry claims that the Christian story is no longer the story of the Earth or the integral story of humankind.[32] He rejects the traditional Christian vision of creation: 'The primary doctrine of the Christian creed, belief in a personal creative principle, became increasingly less important in its functional role.'[33] Berry has blasphemously affirmed that we should 'consider putting the Bible on the shelf for perhaps twenty years, so that we can truly listen to creation.'[34] He has also proposed that 'the only effective program available as our primary guide toward a viable human mode of being is the program offered by the Earth itself'.[35] Berry's tenets stray far from the Church, when he rejects the Christian ideal of being crucified to the world and living only for Christ our Saviour: 'This personal savior orientation has led to an interpersonal devotionalism that quite easily dispenses with earth except as a convenient support for life'.[36] Berry dreamingly sees the world being called to a new post–denominational, even post–Christian, belief system that sees the earth as a mythological living being, as Gaia, Mother Earth, with mankind as her consciousness. Berry's eschatology is false and pernicious:

> Subjective communion with the earth, identification with the cosmic–earth–human process, provides the context in which we now make our spiritual journey... It is no longer simply the journey of the Christian community through history to the heav-

enly Jerusalem. It is the journey of primordial matter through its marvelous sequence of transformations — in the stars, in the earth, in living beings, in human consciousness — toward an ever more complete spiritual–physical intercommunion of the parts with each other, with the whole, and with that numinous presence that has been manifested throughout this entire cosmic–earth–human process.[37]

Berry's approach is more consistent with the views of animistic or shamanistic faiths than anything resembling Christian tradition.

The eco–feminist Rosemary Radford Ruether also proposes a vision which lies far outside the truth of Christianity. The feminist 'theology' she represents is rooted in false principles contrary to any semblance of Catholic doctrine. Ruether often resorts to exalting pagan religions and practices against what she calls the 'patriarchal oppressive' nature of the Catholic Church. In the first place, Ruether believes the Word of God is a collection of myths and that the Bible has to be demythologized, that is, rewritten from the feminist perspective.[38] From early on in her academic career, Ruether had announced her unfavourable disposition towards the Catholic Church and rejected one of its most fundamental beliefs. According to an autobiographical essay, in 1975 she discarded the doctrine of the personal immortality of the soul, the very fulcrum upon which all discipline and doctrine are hinged, during her freshman year at Scripps College.[39] Ruether came to view dogmas not as statements of ontological truth but as useful symbols, and the Church not as a repository of truth, but as a terrible example of what we all are.[40] Towards a 'feminist Christology', she heretically proposes that the 'mythology about Jesus as Messiah or divine Logos, with its traditional masculine imagery', be discarded.[41] Ruether has denied the traditional teachings of the Catholic Church, concerning the sacredness of human life and the family. She has actively supported the mentality of contraception and abortion.[42] Ruether has also espoused pagan worship, with devotion to some female deities like Isis, Athena, and Artemis.[43]

Similarly to be rejected are the antics of the renegade ex–Dominican, and now Episcopalian, Matthew Fox, yet another errant writer on creation spirituality. Fox denies the traditional doctrine of original sin, saying that we do not enter existence as sinful creatures: He claims that we burst into the world as 'Original Blessings'.[44] The only sin Fox recognizes is the sin of dualism, or of seeing people and things as being separate from one another; the only sin is the refusal to see all as one.[45] Fox has written that while 'excess' drug use is not wise, 'intelligent use of drugs' is unquestionably an aid to prayer. Its value, says Fox, is in opening up one's awareness and also as a temporary escape from the worries of the everyday world. He maintains that 'drugs can democratize spirituality, which has for so long been imagined to be in the hands and hearts of the wealthy, leisurely classes.'[46] Fox overturns traditional Christology, insisting that Jesus was not good because He was God, but instead was divine because He was good. This denies the objective divinity of Jesus. Specifically, he writes: 'Jesus is not so much compassionate because he is divine as he is divine because he is compassionate. And did he ... not teach others that they too were ... divine because they are compassionate?'[47]

Like many other ecological ideologists, Fox has drifted off into paganism and witchcraft:

Native American spirituality is a creation–centered tradition, as are the other prepatriarchal religions of the world such as African religions, Celtic religion, and the matrifocal and Wikke traditions that scholars and practitioners like Starhawk are recovering. The contemporary mystical movement known as 'New Age' can also dialogue and create with the creation spiritual tradition.[48]

Some writers have absorbed the alarmist rhetoric and anti–human agenda of secularist environmentalists, in this way blending political and pseudo–religious ideology. One example is the ex–liberation theologian and ex–priest, Leonardo Boff. He applied Marxist dialectics and hermeneutics to 'deep ecology' theory and junk science, and claimed that we should be alarmed by an apparent resource decline as well as population increases. These, he suggested, threatened 'Gaia'—the name for planet Earth, conceived as a superorganism. Boff tends to put the poor and the earth on the same level as being equally oppressed:

> The existence of rich and poor in our societies is in itself a form of ecological aggression. The rich consume too much, wastefully and without thought for the present or future generations; they have set up a technology of death to defend their privileged position, with nuclear and chemical arsenals that could, at worst, bring about biocide, ecocide and even geocide; furthermore, they defend a production system whose inner logic makes it a predator of nature. The poor, victims of the rich, consume less and, in order to survive, live in unhealthy conditions, cut down forests, contaminate waters and soil, kill rare animals and so on. With greater social justice they would be able to operate better environmental justice.[49]

Beyond Boff's philosophically dubious ascription of a type of personhood to the Earth, he ignores the empirical fact that, as prominent economists have pointed out, the price of virtually every commodity (agricultural, mineral, and energy) has fallen steadily throughout the twentieth century.

2.2 *The concept of environment*

The definition of the term 'environment' leads us into the discussion on ecology and its various themes. One can speak of the natural environment, which includes the physical environment with its mineral resources, energy, water, air and so forth. Next there is the plant environment, with its irreplaceable photosynthetic activity: land vegetation, saltwater algae and freshwater algae (in lakes and rivers). Finally to be considered is the animal environment which, together with the plant environment, provides renewable natural resources (food) and also fulfils some ecologically relevant and even irreplaceable activities (for example, insect pollination of flowers).[50] Included in this idea of environment is the notion of a biological chain of processes in dynamic equilibrium, which are important for human beings and their lives.

There are many and various definitions of the environment, and the following steps show how a suitable definition can be constructed starting from a biological approach proceeding to a more human and Christian vision. For A. Auer, the environment is constituted by the whole of our living conditions; therefore, not only 'raw nature,' but also the 'living space created by man.' This same notion of space must take into account the interdependence of man and other living beings.[51] S. Langé continues to develop this distinction between natural and artificial environments. For him, 'the notion of 'environment' today cannot be understood as a natural or primordial fact, but as the result of a historical process.'[52] Further on, in a discussion on the relational aspect, Langé delves deeper into the consideration of the environment in relation to the position that every person has with respect to others, and above all with respect to God. P. Henrici proposes that 'the "natural" environment of human beings is not nature, but rather culture, and therefore a pure and simple "return to nature" is inconceivable. The real human environmental problem consists [...] in the insertion of the cultural (artificial) environment into nature, with all the consequent interactions of these two "environments". The Christian faith [...] plays an important role with regard to the ideal insertion of culture into nature.'[53]

Cardinal Carlo Maria Martini and the Lombardy Episcopal Conference reject a materialist understanding of the notion of environment: 'The human–environment relationship [...] presents complex aspects [...] about which the Christian conscience is called to seek, above all, an initial clarification. Reduced to its most essential terms, it is a question of man's alteration of the dynamic equilibria which guarantee the survival of the biosphere, and therefore of the resources which are necessary for life... Nonetheless, beyond this small reality, environmental crisis can be and is spoken of not only in terms of material resource availability, but also in terms of its meanings and consequent spiritual values.'[54]

Archbishop Renato Martino proposed the following definition at the Rio Conference:

> The word environment itself means 'that which surrounds.' This very definition postulates the

existence of a centre around which the environment exists. That centre is the human being, the only creature in this world who is not only capable of being conscious of itself and of its surroundings, but is gifted with the intelligence to explore, the sagacity to utilize, and is ultimately responsible for its choices and the consequences of those choices. The praiseworthy heightened awareness of the present generation for all components of the environment, and the consequent efforts at preserving and protecting them, rather than weakening the central position of the human being, accentuate its role and responsibilities.[55]

Pope John Paul II, in the Encyclical Letter *Centesimus Annus* (1991), provided an even more complete formula regarding the definition of the environment, which also guards against the danger of cosmocentric neopaganism:

In addition to the irrational destruction of the natural environment, we must also mention the more serious destruction of the *human environment*, something which is by no means receiving the attention it deserves. Although people are rightly worried — though much less than they should be — about preserving the natural habitats of the various animal species threatened with extinction, because they realize that each of these species makes its particular contribution to the balance of nature in general, too little effort is made to *safeguard the moral conditions for an authentic human ecology*. Not only has God given the earth to man, who must use it with respect for the original good purpose for which it was given to him, but man too is God's gift to man. He must therefore respect the natural and moral structure with which he has been endowed. In this context, mention should be made of the

serious problems of modern urbanization, of the
need for urban planning which is concerned with
how people are to live, and of the attention which
should be given to a 'social ecology' of work.[56]

The definitions that we have provided are recent, but there
is a history behind the term 'ecology.' The word 'ecology'
comes from the Greek *ôikos* (= 'house') and *lógos* (= 'speech'
or 'study'). It is the science that studies the relationships of
living beings with one another and with the non–living
environment (soil, water, air, climate). The German naturalist A. von Humboldt (1769–1859) and the French
zoologist G. Saint–Hilaire (1772–1844) were among the
pioneers of ecology. In the field of animal ecology, important studies were conducted by the German naturalist K.
Semper (1832–1893); in the field of plant ecology, major
research was done by the Danish botanist J. E. B. Warming
(1841–1924) and the Swiss botanist A. F. Schimper (1856–
1901).

Ecology makes use of the studies done by two of the
natural sciences, botany and zoology, on the innumerable
life forms that populate the planet, and their classification
and subdivision into species. In the field of ecology, a
'habitat' is the environment in which a given species lives;
a 'niche' is the function that it carries out in a given environment; a 'biotope' is any inhabited physical
environment, in some cases altered as a consequence of its
inhabitants; a 'population' is the whole group of individuals of the same species that live in a given biotope. A
group of populations of various species living in the same
biotope takes the name of 'community,' or 'biocenosis.'
Studying living beings and their environments, ecology
makes use of a pattern or model known as a system
(developed by another 'young' science: 'systems science').

A 'system' is a group of parts, connected to one another, which modify one another over time in an interrelated and orderly way. Thus, an 'ecosystem' is the unit consisting of a community and the environment in which it lives.

An ecosystem is not a closed system, but is connected to other ecosystems via open boundaries ('ecotones'). In this way, the entire natural world is understood as a collection of ecosystems. Each ecosystem behaves as a part, or 'subsystem,' of larger systems. In particular, ecosystems in which a certain community of plant species predominates constitute large ecosystems known as 'biomes' (such as the arctic tundra, steppes, temperate forests, tropical rainforests, African savannas, and deserts), which are divided into subsystems that differ from one another based on the presence of differing communities of animal organisms.

The systematic structuring of the natural world carried out by ecology culminates in the description of the entire natural world, the terraqueous globe and its atmosphere, as a vast and unified system, a system of systems, which is known as the 'terrestrial ecosystem' or 'ecosphere'. Ecology's ambition to interpret all of nature becomes apparent in the formulation of the concept of ecosphere, which starts with the presupposition that individual elements can only be understood if seen as parts of a whole. We cannot limit humanity's environment to solely material elements, because human beings are spiritual and material, and this is the reason that God is humanity's Environment in an eschatological sense. The idea of a universe which is a home for man and woman comes from the Judeo–Christian tradition. The book of Genesis describes the universe as a 'tent'.

Ecology can be defined as the science that deals with organisms in a certain environment, and the processes that connect organisms with places. The distinction must be made, however, between ecology, which is the science, and ecologism, which is an ideology created around the science. In 1962, Rachel Carson published the book *Silent Spring*, which was a sorrowful affirmation on the state of environmental degradation caused by avid and thoughtless human activity, a sombre omen of the death of nature: this date can be marked as the beginning of the modern environmentalist movement.[57] Environmentalism (which has a certain ideological influence in a political sense) was born on April 22, 1970, with 'Earth Day.'[58] Unfortunately, many ideologies of today (such as those of the Greens and Communists) are materialistic, excluding *a priori* from their positions the consideration of God the Creator. There is a great deal of contradiction in the secular position. Abortion is encouraged on the one hand, and the defence of animals is promoted on the other. There is a significant inconsistency in the fact that many democratic countries have the usual abortion law and, at the same time, punish those who, without proven necessity, kill an animal.[59]

Among the Greens, there is a tendency to eliminate the irreducible differences between humans and the rest of creation. In the cultural outlook of the Greens, but also of some other politicians, two primary ideological contaminants can be identified: first and foremost, a renewed philosophical and theological pantheism; second, a materialistic scientism that reduces all sectors of knowledge to the scientific method. These tendencies lead to a reductionism which does not accept openness to the transcendent dimension as a coherent consequence of any non–sectional view. Therefore, in a coherent view, the human environment must include material, biological, intellectual, cultural, moral, and spiritual elements — all in relation to God the Creator. It is therefore necessary to avoid cosmocentrism and exaggerated anthropocentrism. A Christological vision is fundamental in this regard.

In this context, the notion of pollution is not limited to the physical–material or biological realm, but there is also 'pollution' of the information sector with the introduction of viruses into computer programs and information theft. Additionally, there is another type of pollution in the field of social communications. Through social communication means, there is a deception of the mass public in relation to goods, via publicity. Then there is the 'pollution' of family life and of Christian morality by means of pornography:

> Indeed, pornography can militate against the family character of true human sexual expression. The more sexual activity is considered as a continuing frenzied search for personal gratification rather than as an expression of enduring love in marriage, the more pornography can be considered as a factor contributing to the undermining of wholesome family life.[60]

The Internet demands ever more careful precautions against pornography. Other than these forms of cultural and moral pollution, there is also 'pollution' in the intellectual realm, for example in ideologies that contradict a moderate realism, such as idealism, materialism, pragmatism, scientism and nihilism. Along with these ideologies, various political positions arise which counter the right use of reason. Then there is 'pollution' in the realm of faith, deriving from the many ideologies and false notions that oppose the truths revealed and taught by the Church. In a Christian understanding, therefore, pollution cannot be reduced merely to the biological level.

2.3 Pessimism or optimism?

2.3.1 Pessimism

Before World War I, the idea had already begun to spread that the essentially technological progress of civilization would eliminate everything on earth. Others upheld the existence of an intrinsic, self–destructive process of humanity with physical and psychological aspects. In 1969, at a symposium in the Brookhaven Laboratories one participant suggested that the human race has, perhaps, thirty–five years left.[61] Some literary writers such as Huxley, with this *Brave New World*, and Orwell, with his *1984*, depict a scientific and technological society with all of its problems and deviations. Still others, such as C. S. Lewis (who among other things is more optimistic), maintain the following position: along with the ecological degradation, there is an ethical desert; there will be a conquest of man by man himself. That is, man will destroy himself before he destroys nature.

Many of these pessimistic positions lack consequences at the transcendent level. Many think that the end of humanity or of the entire world will occur as a result of merely human or purely physical factors. The idea of Providence in the culmination of history is lacking, and the notion of divine intervention to end history is lacking. The virtue of hope is also lacking.[62] Many times, instead, there is a notion of chance and chaos, sometimes in the form of the law of the jungle (survival of the fittest) of Darwinian or neo–Darwinian origin.

2.3.2 Optimism

While pessimism errs in desperation, optimism errs by presumption and arrogance. The fact that a major part of pessimistic prognoses have been disproved by development itself favours the optimists. First of all, upon the invention of trains and railways, biologists and doctors predicted that the human body would not have been able to sustain such a velocity. Then, when the trip to the moon was planned, again doctors and biologists affirmed that man would not have been able to live in space without the force of gravity, and that after a week at the most he would die!

The position of the optimists is often ingenuous. They believe in a future utopia that can be created by science and technology alone. One example is the idea of freezing the human body after death in order to thaw it later. Often under the influence of Hegel and Teilhard de Chardin, they believe in limitless scientific progress. But, in a limited universe, limitless development is impossible; it is a physical axiom. Thinking that the crisis can be overcome by sprinting forward is like the real–life application of that old scene in the Marx Brothers film where they burn the wood of the train cars to feed the locomotive's furnace.

While many pessimists sustain that destructive power is inherent to humanity, optimists believe that there is a utopian growth force in the universe, an endpoint that will be reached regardless of the crisis. The factor that unites many pessimists and optimists is the negation of a Creator, the negation of any transcendental cause in the universe, and the negation of finality. Both positions are bound to neo–deterministic explanations based on chance. Christian hope is different. A distinction can be made between progressive utopianism (which anticipates an earthly paradise in the future) and conservative utopianism (which wishes to reconstruct a paradise lost in the past). Both are illusory positions. While pessimism often anticipates the final destruction of the universe and of human life as an immanent process within history, optimism anticipates utopia as an immanent process within science; the Christian perspective envisions an end of the universe dependent on a divine decision.[63]

2.3.3 Realistic prognoses

Realism, in this sense, is to be distinguished from the philosophical realism dealt with in greater detail in other works of mine.[64] There is, however, a connection, because realism in looking to the Earth's future regarding ecology must be based on metaphysical realism. In the words of Pope Leo XIII, 'nothing is more useful than to look upon the world as it really is'.[65] Realistic prognoses, in contrast to optimistic ones, are distinguished by the fact that they take into account, as much as possible, some eventual negative consequences in the various interdependent realms and urge a decisive, profound attitude change on humanity's part. The true dangers are those which depend on human beings and their decisions. The current quantitative development plan must be supplemented and completed with a qualitative development plan.

Pessimism is also opposed by the fact that some solutions have been found in the realm of science and technology, such as the efficacious recycling of refuse. S. L. Jaki made an interesting affirmation in this regard: 'We should not forget that many ecological problems which originate in science, or rather in the attitude of the product–acquiring public, can be solved precisely through greater development of the same science.'[66] An example of such a project is research on the possibility of storing radioactive waste formed from the production of nuclear energy. Another approach is research on the production of electrical energy through the process of nuclear fusion, which would be much 'cleaner.' This, however, presupposes humanity's good will. The criterion must not be profit alone, but the common good in a realist perspective.

We must take into account the Christian notion of creation, of original and actual sin, of redemption and the moral life, in order to resolve these ecological issues. There is need for a renewed Christian culture. It is an issue that concerns humanity's relationship with creation (understood as the visible universe of the animal, plant, and mineral realms), with other human beings and, above all, with God. First and foremost, we must ground our thoughts in the teachings of the Catholic Church, which will be the topic of the next chapter.

Notes

1   See D. L. Ray, *Trashing the Planet* (New York: Harper Collins, 1992).

2   See S. L. Jaki, 'Ecology or Ecologism?' in G. B. Marini–Bettòlo (ed.), *Man and his Environment. Tropical Forests and the Conservation of Species* (Vatican City: Pontifical Academy of Sciences, 1994), pp. 271–293.

3   *Ibid.*, p. 276.

4   Cf. *ibid.*.

5   Cf. *ibid.*, p. 277.

6   Broadly, the *Coasian solution* considers that a pollution market, and therefore an efficient allocation mechanism, exists if property rights over environmental assets are well defined and attributed. The Coase theorem, dating from the early sixties, was largely responsible for getting Ronald Coase the Nobel award in Economics, in the early nineties. The Coase theorem formally states that for two economic agents, A and B, when A's actions generate a negative exter-

nality for B, and transaction costs are zero for both parties, it is optimal in terms of social welfare to allow the two agents to negotiate a payment to resolve the issue—either through A's compensating B for the damage A's activity inflicts upon B, or through B's compensating A for the benefits A will forego by discontinuing the activity.

7   See A. Gaspari & V. Pisano, *Dal popolo di Seattle all'ecoterrorismo. Movimenti antiglobalizzazione e radicalismo ambientale* (Milano: 21mo secolo, 2003).

8   See J. Hermand, *Grüne Utopien in Deutschland: Zur Geschichte des ökologischen Bewußtseins* (Frankfurt: 1991), pp. 44–45.

9   W. H. Riehl, *Feld und Wald* (Stuttgart: 1857), p. 52.

10   K. Bergmann, *Agrarromantik und Großstadtfeindschaft* (Meisenheim: 1970), p. 38. There is no satisfactory English counterpart to 'Großstadtfeindschaft', a term which signifies hostility to the cosmopolitanism, internationalism, and cultural tolerance of cities as such. This 'anti–urbanism' is the precise opposite of the careful critique of urbanization worked out by Murray Bookchin in *Urbanization Without Cities* (Montréal: 1992), and *The Limits of the City* (Montréal: 1986).

11   See E. Haeckel, *Generelle Morphologie der Organismen: Allgemeine Grundzüge der organischen Formen–Wissenschaft, mechanisch begründet durch die con Charles Darwin reformierte Descendez–Theorie* (Berlin: Georg Reimer, 1866), I, p. 238 and II, p. 286.

12   D. Gasman, *The Scientific Origins of National Socialism: Social Darwinism in Ernst Haeckel and the German Monist League* (New York: 1971), p. xvii.

13   *Ibid.*, p. 30.

14   See L. Klages, 'Mensch und Erde' in *Sämtliche Werke*, Band 3, (Bonn: 1974), pp. 614–630

15   See M. Zimmerman, *Heidegger's Confrontation with Modernity: Technology, Politics and Art* (Indianapolis: 1990), pp. 242–243.

16   E. Lehmann, *Biologischer Wille. Wege und Ziele biologischer Arbeit im neuen Reich* (München: 1934), pp. 10–11.

17   B. Commoner, *The Closing Circle: Nature, Man and Technology* (New York: Alfred A. Knopf, 1971), p. 29.

18   See J. Hemleben, *Rudolf Steiner und Ernst Haeckel* (Stuttgart: 1965 and K. Ballmer, *Rudolf Steiner und Ernst Haeckel* (Hamburg: 1929). Theosophy is a doctrine of pseudo–religious philosophy and metaphysics started by Helena Petrovna Blavatsky (1831–1891). In this context, theosophy holds that all religions are attempts by the 'Spiritual Hierarchy' to help humanity in evolving to greater perfection, and that each religion therefore has a portion of the truth. Theosophy seems to a clear precursor of New Age.

19   R. Steiner, *Lecture*, 9 June 1904.

20   P. Teilhard de Chardin, *The Phenomenon of Man* (New York: Harper & Row, 1961), p. 301.

21   *Ibid.*, pp. 294, 310.

22   P. Teilhard de Chardin, Letter cited in Philippe de la Trinité, *Rome et Teilhard de Chardin* (Paris: Fayard, 1964), p. 168.

23   P. Teilhard de Chardin, *The Heart of Matter* (London : Collins, 1978), pp. 52–54.

24   P. Teilhard de Chardin, Letter cited in Philippe de la Trinité, *Rome et Teilhard de Chardin* (Paris: Fayard, 1964), p. 168.

25   P. Teilhard de Chardin, 'Contingence de l'univers et goût humain de survivre' (1953) unpublished essay, p. 4.

26   See C. Cuénot, *Teilhard de Chardin* (London: Burns Oates, 1965), p. 293.

27   P. Teilhard de Chardin, *Hymn of the Universe* (London: Collins, 1965), p. 133.

28   P. Teilhard de Chardin, *Letters from a Traveller* (New York: Harper, 1962), p. 269.

29   See P. Teilhard de Chardin, *Christianity and Evolution* (New York: Harper, 1971), pp. 216f..

30   P. Teilhard de Chardin, *The Divine Milieu* (New York:

Harper & Row, 1960), p. 122.

31 See *L'Osservatore Romano* (1 July 1962 , n.148), which refers
to the *Monitum* (dated 30 June 1962 in *AAS* 54 (1962), p. 166)
directed at the errors of P. Teilhard de Chardin. The text
runs as follows: 'Several works of Fr. Pierre Teilhard de
Chardin, some of which were posthumously published, are
being edited and are gaining a good deal of success.
Prescinding from a judgment about those points that
concern the positive sciences, it is sufficiently clear that the
above mentioned works abound in such ambiguities, and
indeed even serious errors, as to offend Catholic doctrine.
For this reason, the eminent and most revered Fathers of the
Supreme Sacred Congregation of the Holy Office exhort all
Ordinaries, as well as Superiors of Religious institutes,
rectors of seminaries and presidents of universities, effectively to protect the minds, particularly of the youth, against
the dangers presented by the works of Fr. Teilhard de
Chardin and of his followers.' A year later, in 1963, the
Vicariate of Rome required that Catholic booksellers in
Rome should withdraw from circulation the works of Teil-
hard de Chardin, along with any other books which
supported his views. In 1967, the Apostolic Delegation in
Washington affirmed that the *Monitum* was still in place. In
1981, this same affirmation was repeated, this time by the
Vatican itself. The following is the text of the 1981 statement
(see *L'Osservatore Romano*, 20 July 1981): 'The letter sent by
the Cardinal Secretary of State to His Excellency Mgr.
Poupard on the occasion of the centenary of the birth of Fr.
Teilhard de Chardin has been interpreted in a certain section
of the press as a revision of previous stands taken by the
Holy See in regard to this author, and in particular of the
Monitum of the Holy Office of 30 June 1962, which pointed
out that the work of the author contained ambiguities and
grave doctrinal errors. The question has been asked whether
such an interpretation is well founded. After having
consulted the Cardinal Secretary of State and the Cardinal

Prefect of the Sacred Congregation for the Doctrine of the Faith, which, by order of the Holy Father, had been duly consulted beforehand, about the letter in question, we are in a position to reply in the negative. Far from being a revision of the previous stands of the Holy See, Cardinal Casaroli's letter expresses reservation in various passages — and these reservations have been passed over in silence by certain newspapers — reservations which refer precisely to the judgment given in the *Monitum* of June 1962, even though this document is not explicitly mentioned.'

32   T. Berry, *The New Story*. Teilhard Studies n° 1 (Chambersburg, PA: Anima Press, 1978).

33   *Ibid.*, p. 2.

34   M. Hope & J. Young, 'A Prophetic Voice: Thomas Berry' in *Trumpeter* 11/1 (1994), p. 16.

35   T. Berry, *The Great Work* (New York: Random House, 1999), p. 71.

36   T. Berry, 'The Spirituality of the Earth' in C. Birch, W. Eaken & J. B. McDaniel (eds.), *Liberating Life: Contemporary Approaches in Ecological Theology* (New York: Orbis Books, 1990), pp. 151–158.

37   T. Berry, 'The Spirituality of the Earth' in C. Birch, W. Eaken and J. B. McDaniel (eds.), *Liberating Life: Contemporary Approaches in Ecological Theology* (New York: Orbis Books, 1990), pp. 151–158.

38   See C. R. Ferreira, *The Feminist Agenda Within the Catholic Church* (Toronto: Life Ethics Centre, 1987), p. 4.

39   See Ruether's autobiographical essay 'Beginnings: An Intellectual Autobiography', in G. Baum, (ed.), *Journeys: The Impact of Personal Experience on Religious Thought* (New York: Paulist Press, 1975), p. 34.

40   Cf. *ibid.*.

41   R. R. Ruether, *Sexism and God–Talk* (Boston: Beacon Press, 1983), p. 137

42   R. R. Ruether, 'Women, Sexuality, Ecology, and the Church', in *Conscience* (Spring/Summer 1993), pp. 6, 10.

43   R. R. Ruether, 'The Hideous Error of Women Priests' in *Crying in the Wilderness Newsletter* (Autumn 1992), p. 4.

44   R. R. Ruether, *Mary–The Feminine Face of the Church* (Philadelphia: Westminster, 1979), pp. 13–17.

45   See M. Fox, *Original Blessing: A Primer in Creation Spirituality* (Santa Fe, NM: Bear & Company, 1983), pp. 47, 49.

46   M. Fox, *On Becoming a Musical, Mystical Bear: Spirituality American Style* (New York, NY: Paulist Press, 1976), pp. 125–127.

47   M. Fox, *A Spirituality Named Compassion and the Healing of the Global Village, Humpty Dumpty and Us* (Minneapolis, MN: Winston Press, 1979), p. 34.

48   M. Fox, *Original Blessing*, p. 16. Miriam Simos (Starhawk) is a practicing witch on the staff of Matthew Fox's Institute for Culture and Creation Spirituality (ICCS).

49   L. Boff & V. Elizondo, 'Ecology and Poverty: Cry of the Earth, Cry of the Poor' in *Concilium* 5/1995, p. xi. See also L. Boff, *Ecology and Liberation, a New Paradigm* (Maryknoll: Orbis, 1995); Idem, *Cry of the Earth, Cry of the Poor* (Maryknoll: Orbis, 1997).

50   See P. C. Beltrão, *Ecologia umana e valori etico–religiosi*, (Rome: Editrice Pontificia Università Gregoriana, 1985), p. 33.

51   See A. Auer, *Etica dell'ambiente*, Queriniana, Brescia 1988, p. 14: 'rozza natura'; 'spazio vitale plasmato dall'uomo.'

52   S. Langé, 'Ecologia e tutela dell'ambiente costruito', in A. Caprioli & L. Vaccaro, *Questione ecologica e coscienza cristiana*, (Brescia: Morcelliana, 1988), p. 57: 'la nozione "ambiente" oggi non può essere concepita come dato naturale o primordiale, ma come esito di un processo storico.'

53   P. Henrici, 'Essere umano e natura nell'era della tecnologia,' in P. C. Beltrão, *Ecologia umana e valori etico–religiosi*, p. 76: 'l'ambiente "naturale" dell'essere umano non è la natura,

bensì la cultura, e pertanto un puro e semplice 'ritorno alla natura' è inconcepibile. Il vero problema ecologico umano consiste [...] nell'inserimento dell'ambiente culturale (artificiale) nella natura, con tutte le interazioni di questi due "ambienti." La fede cristiana [...] gioca un ruolo importante riguardo all'inserimento ideale della cultura nella natura.'

54   Lombardy Episcopal Conference, *La questione ambientale* (Milan: Centro Ambrosiano, 1988), p. 15. 'Il rapporto uomo–ambiente [...] presenta aspetti complessi [...] di fronte ai quali la coscienza cristiana è chiamata a provocare anzitutto un chiarimento di principio. Ridotta ai suoi termini più essenziali essa è la questione dell'alterazione, a opera dell'uomo, di quegli equilibri dinamici che garantiscono la sopravvivenza della biosfera e, dunque, anche della risorse indispensabili alla vita... Tuttavia, al di là di questa determinazione minima, si può parlare, e di fatto si parla, di crisi dell'ambiente, non soltanto sotto il profilo delle sue disponibilità materiali, ma anche sotto il profilo dei suoi significati e dei conseguenti valori spirituali.'

55   Archbishop Renato Martino, *Statement to the United Nations Conference on Environment and Development*, Rio de Janeiro (4 June 1992), in Appendix 3, p. 294 below.

56   Pope John Paul II, Encyclical Letter *Centesimus Annus* (1991), 38.

57   See R. Carson, *Silent Spring* (Boston: Houghton Mifflin, 1962).

58   See C. M. Murphy, *At Home on Earth, Foundations for a Catholic Ethic of the Environment* (New York: Crossroad, 1989), p. 30.

59   See M. Gargantini, 'I cristiani e le tematiche ambientaliste,' in A. Caprioli & L. Vaccaro, *Questione ecologica e coscienza cristiana*, p. 93: 'significativo il caso del Canada che 'da buon paese modernizzato ha la solita legge abortista' e multa...chi, senza provate necessità, uccide un animale...'

60   Pontifical Council for Social Communications, *Pornography and Violence in the Communications Media: A Pastoral Response*

(7 May 1989), 16.

61 According to J. B. Cobb, *Is It Too Late? A Theology of Ecology* (Beverly Hills, Ca.: Bruce, 1972), p. 13.

62 See Chapter 4, subsection 4.3.10 below.

63 See Chapter 4, subsection 4.3.10 below.

64 Cfr. P. Haffner, *The Mystery of Reason* (Leominster: Gracewing, 2001), pp. 12–19.

65 Pope Leo XIII, *Rerum Novarum*, 18.

66 S. L. Jaki, 'Intervento al Meeting per l'amicizia fra i popoli del 1988,' in Meeting '88. *Cercatori di Infinito. Costruttori di Storia* (Rimini: 1989), p. 204: 'Non dovremmo dimenticare che molti problemi ecologici che hanno origine nella scienza, o meglio nell'atteggiamento del pubblico che acquista i prodotti, possono trovare la loro soluzione proprio in un maggior sviluppo della medesima scienza.'

3

Christian teaching on ecology

The essential meaning of this 'kingship' and 'dominion' of man over the visible world, which the Creator Himself gave man for his task, consists in the priority of ethics over technology, in the primacy of the person over things, and in the superiority of spirit over matter.

Pope John Paul II, Redemptor hominis

3.1 Papal teaching

There is a long tradition of Pontifical teaching on the goodness and beauty of creation going all the way back to Pope Clement I, in the first century.

> The heavens, revolving under His government, are subject to Him in peace. Day and night run the course appointed by Him, in no wise hindering each other. The sun and moon, with the companies of the stars, roll on in harmony according to His command, within their prescribed limits, and without any deviation. The fruitful earth, according to His will, brings forth food in abundance, at the proper seasons, for man and beast and all the living

beings upon it, never hesitating, nor changing any of the ordinances which He has fixed. The unsearchable places of abysses, and the indescribable arrangements of the lower world, are restrained by the same laws. The vast unmeasurable sea, gathered together by His working into various basins, never passes beyond the bounds placed around it, but does as He has commanded. The ocean, impassable to man and the worlds beyond it, are regulated by the same enactments of the Lord. The seasons of spring, summer, autumn, and winter, peacefully give way to one another. The winds in their several quarters fulfil, at the proper time, their service without hindrance. The ever–flowing fountains, formed both for enjoyment and health, furnish without fail their breasts for the life of men. The very smallest of living beings meet together in peace and concord. All these the great Creator and Lord of all has appointed to exist in peace and harmony.[1]

Our own analysis of Papal teaching on strictly ecological issues starts with Pope Pius XII.

3.1.1 Pius XII

Addressing a group of Italian agriculturists in the now–distant year of 1946, Pius XII showed himself a defender of a true 'Christian ecology.' In speaking about the Italian countryside, he cited Pliny in extolling its lively and perennial wholesomeness, fertile fields, luminous hills, large herds, shady forests, and fruitful vines and olive trees.[2] The Pope praised the fact that the farmers lived in continual contact with nature, that their lives were lived in areas still far removed from the excesses of an artificial society, and were entirely directed toward drawing up from the depths of the soil, under the sun of the divine Father, the abundant treasures that His hand had hidden.[3] Pius XII also indicated the negative effects that sin brought into the world. Before sin, God had given the land to man so that he could cultivate it, as the most beautiful and honourable work in the natural order. Continuing the sinful action of our first parents, the actual sins of humanity have continually worsened the curse upon the earth: 'Repeatedly struck by all of the calamities, floods, terrestrial cataclysms, pestilences and devastating wars, the soil, in some places deserted, sterile, and unhealthy, ... refused to freely bestow its treasures upon man. The earth is greatly wounded, greatly afflicted. Bent over her, not as a slave over the field, but as a clinician at the patient's bedside, the cultivator lavishes her with attention and love.'[4] However, as the Pope added, love must be accompanied by vast and broad awareness.[5]

Another message by Pope Pius XII exemplifies how the Church anticipated the concerns of society regarding the environment. Far from being behind the secular world on this issue, the Church was the first to deal with it, but in a faith context:

> We have observed, without anxiety or trepidation, the recent developments which, after several power facilities (of electronuclear energy production), have brought the first attempt to move a ship through nuclear transformation energy to a good conclusion, finally using this force in the service of man and not for his destruction ... With equal confidence and hope we are following many research efforts which are studying the effects that various types of currently useable radiation have on plants, on their possibility of conservation ... Nonetheless, regarding what research can do in the coveted realm of life, we must once again warn against the

dangers, which genetics foresees as possible, that arise when the mystery at the basis of all living things is tampered with by imprudent interventions or by a violent habitat change, such as an increase of radioactivity levels beyond an as–yet unknown threshold of biological safety.[6]

3.1.2 Paul VI

Pope Paul VI further developed the Church's outlook concerning ecology. The first part of his Pontificate saw the conclusion of the Second Vatican Council, which made a very important basic consideration on this issue:

> For man, created to God's image, received a mandate to subject to himself the earth and all it contains, and to govern the world with justice and holiness; a mandate to relate himself and the totality of things to Him Who was to be acknowledged as the Lord and Creator of all. Thus, by the subjection of all things to man, the name of God would be wonderful in all the earth.[7]

This passage speaks about the recapitulation of earthly things in Christ and affirms that the submission of the earth to man is not arbitrary, but must be carried out in accordance with the natural and revealed law. Through man's work, creation is brought to fulfilment. Paul VI more explicitly addressed the ecological question, even in a much broader sense, as related to the moral problem:

> Today people are concerned with ecology, that is, with the purification of the physical environment in which our human lives are carried out: why are we not also concerned with a moral ecology in which human beings live as human beings and children of God?[8]

In one of his important social documents, the Pope addressed the communal aspect of this issue:

> While the horizon of man is thus being modified according to the images that are chosen for him, another transformation is making itself felt, one which is the dramatic and unexpected consequence of human activity. Man is suddenly becoming aware that by an ill–considered exploitation of nature he risks destroying it and becoming in his turn the victim of this degradation. Not only is the material environment becoming a permanent menace — pollution and refuse, new illness and absolute destructive capacity — but the human framework is no longer under man's control, thus creating an environment for tomorrow which may well be intolerable. This is a wide–ranging social problem which concerns the entire human family. The Christian must turn to these new perceptions in order to take on responsibility, together with the rest of men, for a destiny which from now on is shared by all.[9]

Paul VI also discussed concrete environmental questions:

> But how can we ignore the imbalances caused in the biosphere by the disorderly exploitation of the physical reserves of the planet, even for the purpose of producing something useful, such as the wasting of natural resources that cannot be renewed; pollution of the earth, water, the air and space, with the resulting attacks on plant and animal life? All that contributes to the impoverishment and deterioration of man's environment to the extent, it is said, of threatening his own survival... Just as the demographic problem is not solved by unduly limiting access to life, so the problem of the environment cannot be tackled with technical measures alone. A

mentality change is necessary... But all technical measures would remain ineffectual if they were not accompanied by awareness of the necessity of a radical change of mentality. How can we fail to recall here the imperishable example of St. Francis of Assisi and to mention the great Christian contemplative Orders, which offer the testimony of an inner harmony achieved in the framework of trusting communion with the rhythms and laws of nature?[10]

The notion of ecological equilibrium lies at the basis of all sound environmental reflection, as the Pope suggested to the Pontifical Academy of Sciences:

You feel deeply within you the solidarity that binds you to mankind today and in the future, and that is why you adopt an attitude which is that of the ever serious scientist, the attitude of one who — as we had the opportunity to stress during our meeting last year — 'must honestly consider the question of the earthly future of mankind and, as a responsible person, contribute to prepare it, preserve it and eliminate risks'. The subject chosen for the present Week reflects this concern in an evident way. With regard to the harmful agents which threaten plants, the fruits of which constitute directly or indirectly the main source of subsistence for the human being, protection is carried out today thanks above all to synthetic chemical products. But the latter are causing more and more serious concern, owing to their possible long–term toxic effects on man, and owing, too, to the changes they bring to the natural environment, with the consequent disturbances of the ecological balance.[11]

All of these ecological issues demand a new solidarity involving future generations, and a generous sense of sharing between rich and poor.[12]

3.1.3 *John Paul II*

Already with his first encyclical, Pope John Paul II emphasized the principles that must form the foundation of any sound ecological discussion:

> The man of today seems ever to be under threat from what he produces ... This seems to make up the main chapter of the drama of present–day human existence in its broadest and universal dimension ... This state of menace for man from what he produces shows itself in various directions and various degrees of intensity. We seem to be increasingly aware of the fact that the exploitation of the earth, the planet on which we are living, demands rational and honest planning. At the same time, exploitation of the earth not only for industrial but also for military purposes and the uncontrolled development of technology outside the framework of a long–range authentically humanistic plan often bring with them a threat to man's natural environment, alienate him in his relations with nature and remove him from nature. Man often seems to see no other meaning in his natural environment than what serves for immediate use and consumption. Yet it was the Creator's will that man should communicate with nature as an intelligent and noble 'master' and 'guardian,' and not as a heedless 'exploiter' and 'destroyer.'[13]

The Pope pronounced against pragmatism both in ecology and in other areas of human life:

> The essential meaning of this 'kingship' and

'dominion' of man over the visible world, which the Creator himself gave man for his task, consists in the priority of ethics over technology, in the primacy of the person over things, and in the superiority of spirit over matter.[14]

That same year, the Pope repeated this passage of *Redemptor hominis*, and grounded this essential meaning of human sovereignty, in contrast with neo–pagan cosmocentrism:

This threefold superiority is maintained to the extent to which the sense of the transcendence of man over the world and of God over man, is preserved ... Man must emerge victorious from this drama which threatens to degenerate into a tragedy, and must rediscover his true kingship over the world and his full dominion over the things he produces.[15]

During Pope John Paul II's Pontificate, Saint Francis of Assisi was declared the patron saint of ecology. This declaration recognizes that he must be

rightly counted among the saints and illustrious men who had a unique veneration for nature, considered a magnificent gift from God to humanity ... In fact, he had a unique perception of all of the Creator's works, and nearly superbly inspired he composed the beautiful Canticle of the Creatures, through which — particularly through brother sun, sister moon and the stars — he gave due praise, honour, glory and every blessing to the highest, omnipotent and good Lord.[16]

The Pope also addressed the question of resources in relation to the ecological question, in the context of considering the environment an inheritance. He also cited his own work experience:

> In the course of history, man has developed the forms of energy that he needed, passing from the discovery of fire to ever richer forms of energy, and arriving finally at nuclear energy, which is staggering from so many points of view. At the same time, the progress of industrialization has given rise, especially in recent times, to ever increasing consumption, to such an extent that certain natural resources are now becoming exhausted. Our civilization — above all its scientists and technicians — must look for new methods in order to use the energy sources that Divine Providence has put at the disposal of man. It is necessary, furthermore, that governments themselves should pursue a unified energy policy, so that the energy produced in one region can be used in other regions ... I myself have seen the harm done to the beauty of nature by industrial installations which could have been placed elsewhere or planned differently. Above all, I have had personal experience of the sufferings of coal miners, whose lungs are impregnated with the dust that poisons the mine tunnels. I hope and trust that, in the name of human rights and for the improvement of the quality of life, new and effective measures have already been adopted for the utilization of conventional sources of energy, and that in this way we will no longer have to see jeopardized not only the natural environment, but also workers and populations ... Energy is a universal good that Divine Providence has put in the service of man, of all men, to whatever part of the world they may belong, and we must think also of the men of the future, because the Creator entrusted the earth and the multiplication of its

inhabitants to man's responsibility. I think that it can be considered a duty of justice and charity to make a resolute and persevering effort to husband energy resources and respect nature, so that not only humanity as a whole today may benefit, but also the generations to come.[17]

A novelty offered by the Pope was his declaration on the treatment of animals:

It is certain that animals are at the service of man and can hence be the object of experimentation. Nevertheless, they must be treated as creatures of God which are destined to serve man's good, but not to be abused by him. Hence the diminution of experimentation on animals, which has progressively been made ever less necessary, corresponds to the plan and well–being of all creation.[18]

At the same time, the Pope rejected the grave error of cosmocentrism, which places animals on the same level as the human person: 'in the name of an idea inspired by egocentrism and biocentrism it is being proposed that the ontological and axiological difference between men and other living beings be eliminated, since the biosphere is considered a biotic unity of undifferentiated value. Thus *man's superior responsibility* is eliminated in favour of an egalitarian consideration of the 'dignity' of all living beings.'[19]

Pope John Paul II also indicated St Francis as an example concerning the treatment of animals in this context:

Saint Francis stands before us also as an example of unfaltering gentleness and sincere love toward non–reasoning beings, which are part of creation. In him, the harmony depicted by the suggestive words

of the first pages of the Bible reechoes: 'The Lord
God then took the man and settled him in the
garden of Eden, to cultivate and care for it' (Gn
2:15), and 'brought' the animals 'to the man, to see
what he would call them' (Gn 2:19).[20]

The Pope noted that, in Saint Francis, an anticipation of the
peace foretold in Sacred Scripture can be glimpsed, when
'the wolf shall be a guest of the lamb, and the leopard shall
lie down with the kid; The calf and the young lion shall
browse together, with a little child to guide them' (Is 11:6).
He looked upon creation with the eyes of someone who
recognizes the marvellous work of the hand of God in it.
His voice, his gaze, and his attentive care not only toward
people but also toward animals and nature in general, are
a faithful echo of the love with which God pronounced the
'*fiat*' in the beginning, which brought them into existence.
In the 'Canticle of the Creatures,' how can one not feel the
stirring of something of the transcendent joy of God the
Creator, of Whom it is written that He 'looked at everything He had made, and He found it very good (Gn 1:31)?
Is this perhaps not the explanation of the sweet appella-
tives of 'brother' and 'sister' with which St. Francis
addresses every created being? We are also called to a
similar outlook. Created in the image and likeness of God,
we must make Him present among the creatures 'as intelligent and noble stewards and lords' of nature, and 'not as
unscrupulous exploiters and destroyers.'[21]

Another fundamental key for understanding the
theology of the environment is the concept of beauty. The
Pope has effectively taught that there is a hierarchy of
beauty in creation, based on metaphysics, which is manifested at the natural or artificial aesthetic level, and also at
the functional level:

> Educating to respect for animals and, in general, for
> the harmony of creation has, among other things, a
> beneficial effect on human beings as such, contrib-
> uting to the development of sentiments of balance,
> moderation, and nobility, and habituating the mind
> to see 'from the greatness and the beauty of created
> things' the transcendent beauty and greatness of
> their Author (cf. Ws 13:5).[22]

The beauty of creation was illustrated on one of the trips of
Pope John Paul II to his native Poland:

> As I make my way across Poland, from the Baltic,
> through Great Poland, Mazovia, Warmia and
> Masuria, and then the eastern regions — from the
> region of Bialystok to that of Zamosc — I contem-
> plate the beauty of this, my native country, and I am
> reminded of this particular aspect of the saving
> mission of the Son of God. Here, the blue of the sky,
> the green of the woods and fields, the silver of the
> lakes and rivers, all seem to speak with exceptional
> power. Here the song of the birds sounds so very
> familiar, so Polish. And all this testifies to the love
> of the Creator, the life–giving power of his Spirit
> and the redemption accomplished by the Son for
> man and for the world. All these creatures bespeak
> their holiness and dignity, regained when the One
> who was 'the firstborn of all creation' took flesh
> from the Virgin Mary.[23]

The Pope has proposed that man and his world (our Earth,
which upon the first space voyage appeared as a green and
blue 'star') must be safeguarded and aided in progressing.
This means using life with caution, including animal life
and all animate and inanimate components of nature. The
earth, in the faith perspective, is not a reserve to be
endlessly plundered. It is part of the mystery of creation

which must not merely be utilized, but must inspire awe and reverence.[24]

The cosmic dimension of the ecological debate was explained by the Pope, insisting on the specificity of the Incarnation:

> The Incarnation of God the Son signifies the taking up into unity with God not only of human nature, but in this human nature, in a sense, of everything that is 'flesh': the whole of humanity, the entire visible and material world. The Incarnation, then, also has a cosmic significance, a cosmic dimension. The 'first–born of all creation,' becoming incarnate in the individual humanity of Christ, unites himself in some way with the entire reality of man, which is also 'flesh' — and in this reality with all 'flesh,' with the whole of creation.[25]

The Pope considered the notion of the quality of life, but in a human and Christian perspective, as opposed to a merely materialistic and hedonistic understanding:

> The harmonious relationship between man and nature is a fundamental element of civilization, and it is easy to grasp all the contribution that science can bring in this field of *ecology*, in the form of defence against violent alterations of the environmental and of growth in the quality of life through the humanisation of nature.[26]

John Paul II declared that science must always be oriented toward the good of humanity, and not reduced to mere utility as in the pragmatist understanding. In particular, he states that a perspective which only takes profit into consideration has caused damage to the environment.[27]

In other words, there are limits to human sovereignty, as the Pope affirmed:

Nor can the moral character of development exclude respect for the beings which constitute the natural world ... Such realities also demand respect, by virtue of a threefold consideration which it is useful to reflect upon carefully. The first consideration is the appropriateness of acquiring a growing awareness of the fact that one cannot use with impunity the different categories of beings, whether living or inanimate — animals, plants, the natural elements — simply as one wishes, according to one s own economic needs. On the contrary, one must take into account the nature of each being and of its mutual connection in an ordered system, which is precisely the cosmos. The second consideration is based on the realization — which is perhaps more urgent — that natural resources are limited; some are not, as it is said, renewable. Using them as if they were inexhaustible, with absolute dominion, seriously endangers their availability not only for the present generation but above all for generations to come. The third consideration refers directly to the consequences of a certain type of development on the quality of life in the industrialized zones. We all know that the direct or indirect result of industrialization is, ever more frequently, the pollution of the environment, with serious consequences for the health of the population ... The dominion granted to man by the Creator is not an absolute power, nor can one speak of a freedom to 'use and misuse,' or to dispose of things as one pleases. The limitation imposed from the beginning by the Creator himself and expressed symbolically by the prohibition not to 'eat of the fruit of the tree' (cf. Gn 2:16–17) shows clearly enough that, when it comes to the natural world, we are subject not only to biological laws but also to moral ones, which cannot be violated with impunity.[28]

The affirmation of the reality of the universe and its laws as a basis for ecological reflection is significant here. The notion of law must also be understood on various levels: the laws of physics and biology, juridical laws, and moral laws. The Christian understanding turns out to be in conflict with the positivist notion of law, because law transcends the visible realm. The debate on the notions of ecological equilibrium and resource renewability must also be taken into account. Water can be considered a renewable resource, whereas minerals are non–renewable. There are some limits on renewability, however, leading to a distinction between partial and total renewability.

The primary Pontifical document concerning the environment is the Message for the World Day of Peace on 1 January 1990, *Peace with God the Creator: Peace with all of Creation*.[29] It stated that collective egoism, insecurity, and a lack of due respect for nature threaten peace in the world today. It is therefore necessary to encourage the development of an ecological awareness. There are many ethical values correlated to the environmental question which have to do with peace. In order to resolve, or at least attempt to resolve, the environmental question, it is necessary to find solutions that have a morally coherent world view at their foundation. Christians find this morally coherent view by drawing on Revelation. Adam and Eve were called by God to participate in the fulfilment of His plan for creation, enacting an ordered relationship between them and all of creation. But man fell into sin, going against the design of God the Creator and leading to the subjection of creation to frailty. All of creation was reconciled in Christ.

When man distances himself from the plan of his Creator, he causes disorder which reverberates through all of creation. This fact evidences the relationship between human activity and the whole of creation. All people, not just Christians, witness the devastation caused by those who do not recognize the order and harmony of the cosmos. But, to remedy this devastation, it is necessary to identify the cause of it all: a profound moral crisis. Therefore, the environmental crisis has moral roots. The well–being of future generations and the whole of the ecosystem must be taken into account in the management of scientific and technological progress.

Today everything is done in the name of progress. This leads to a lack of respect for the life and dignity of human beings, whereas progress should be measured precisely in terms of that life and dignity. The earth is a common heritage. Precisely for this reason, it is not just for a small number of people to manage, especially when ecological instability is caused primarily by those few people. The right to a safe environment should be included in a future Charter of Human Rights. There is a need for international coordination regarding ecological questions. This must not lessen the responsibility of the individual States, which must monitor their own territory and pay greater attention to the most vulnerable sectors of the society.

There is an urgent moral need for a new solidarity among States, both so that developing nations do not repeat the same errors made by industrialized nations, and so that peaceful relations can be fostered among the States themselves. A sound ecosystemic equilibrium will never be achieved without addressing the structural forms of poverty. Any war causes incalculable ecological damage. In order to find a solution, it is necessary to take a look at the society's lifestyle, educating in ecological responsibility toward oneself, others, and the environment. Creation also has aesthetic value. Everyone is responsible for order in the world, but Christians are so in virtue of their faith in God the Creator. Respect for life and the dignity of the human person necessarily includes respect and care for creation. Saint Francis of Assisi shows us how peace with God brings peace with creation and with all people.

In another great social encyclical, John Paul II spoke about responsibility for creation as an original gift from God to man:

> Equally worrying is *the ecological question* which accompanies the problem of consumerism and which is closely connected to it. In his desire to have and to enjoy rather than to be and to grow, man consumes the resources of the earth and his own life in an excessive and disordered way. At the root of the senseless destruction of the natural environment lies an anthropological error, which unfortunately is widespread in our day. Man, who discovers his capacity to transform and in a certain sense create the world through his own work, forgets that this is always based on God's prior and original gift of the things that are. Man thinks that he can make arbitrary use of the earth, subjecting it without restraint to his will, as though it did not have its own requisites and a prior God–given purpose, which man can indeed develop but must not betray. Instead of carrying out his role as a co–operator with God in the work of creation, man sets himself up in place of God and thus ends up provoking a rebellion on the part of nature, which is more tyrannized than governed by him.[30]

Pope John Paul II often reaffirmed that the ecological problem is linked to a broader ethical and moral crisis:

> The goods of the earth, which in the divine plan must be a common inheritance, sometimes risk being monopolized by a small number of people. They are used exclusively for the good of some, who often tamper with and occasionally destroy them, thus causing damage to the whole of humanity. It is necessary to stop the race toward the egoistic use of the goods of the earth. Their destruction and alteration must be halted, because we all suffer the negative consequences of thoughtless ecological decisions.[31]

The *Catechism of the Catholic Church*, promulgated by John Paul II in 1992, often speaks of creation, and in particular of the environmental question. Creation was willed by God as a gift to man, as 'an inheritance destined for and entrusted to him.'[32] Human beings are like the priests or stewards of creation because God created everything for man, but man was created to love and serve God and to offer Him all of creation.[33] In God's plan, man and woman are called to be 'stewards of God,' called to participate in Divine Providence toward other creatures. This sovereignty, however, must not be 'an arbitrary and destructive domination.'[34] The mastery of the world that God conceded to man in the beginning must be realized first and foremost in man himself, as self–mastery.[35]

Creation is thus hierarchical, in contrast to the vision proposed by neo–pagan cosmocentrism:

> The *hierarchy of creatures* is expressed by the order of the 'six days,' from the less perfect to the more perfect. God loves all his creatures and takes care of each one, even the sparrow. Nevertheless, Jesus

said: 'You are of more value than many sparrows'
[Lk 12:6–7], or again: 'Of how much more value is a
man than a sheep!' [Mt 12:12][36]

Respect for the integrity of creation is included in the
context of the catechesis on the seventh commandment:

The seventh commandment enjoins respect for the
integrity of creation. Animals, like plants and inanimate beings, are by nature destined for the
common good of past, present, and future humanity. Use of the mineral, vegetable, and animal
resources of the universe cannot be divorced from
respect for moral imperatives. Man's dominion
over inanimate and other living beings granted by
the Creator is not absolute; it is limited by concern
for the quality of life of his neighbour, including
generations to come; it requires a religious respect
for the integrity of creation.[37]

The Catechism also affirms the importance of respect for
animals, which is not to be confused however with so–
called animal rights:

Animals are God's creatures. He surrounds them
with his providential care. By their mere existence
they bless him and give him glory. Thus men owe
them kindness. We should recall the gentleness
with which saints like St. Francis of Assisi or St.
Philip Neri treated animals. God entrusted animals
to the stewardship of those whom he created in his
own image. Hence it is legitimate to use animals for
good and clothing. They may be domesticated to
help man in his work and leisure. Medical and
scientific experimentation on animals is a morally
acceptable practice if it remains within reasonable
limits and contributes to caring for or saving human
lives. It is contrary to human dignity to cause

animals to suffer or die needlessly. It is likewise unworthy to spend money on them that should as a priority go to the relief of human misery. One can love animals; one should not direct to them the affection due only to persons.[38]

The moral basis for ecological action was treated by the Pope in his encyclical *Veritatis splendor*. Here, man is seen as the steward of creation: *"The exercise of dominion over the world* represents a great and responsible task for man, one which involves his freedom in obedience to the Creator's command: 'Fill the earth and subdue it' (Gn 1:28)."[39] The document specifies that man must be the master of himself, not just of the world: 'Not only the world, however, but also *man himself* has been *entrusted to his own care and responsibility.* God left man "in the power of his own counsel" (Sir 15:14), that he might seek his Creator and freely attain perfection.'[40]

This self–mastery translates into solidarity with others in order to overcome the pragmatism which is so widely spread today:

It is a serious abuse and an offence against human solidarity when industrial enterprises in the richer countries profit from the economic and legislative weaknesses of poorer countries to locate production plants or accumulate waste which will have a degrading effect on the environment and on people's health ... Mere utilitarian considerations or an aesthetical approach to nature cannot be a sufficient basis for a genuine education in ecology. We must all learn *to approach the environmental question with solid ethical convictions* involving responsibility, self–control, justice and fraternal love.[41]

The ecological debate can be considered in the light of bioethics, and the basis for this was laid out in the encyclical *Evangelium vitae*. This document notes the positive developments in ecology: 'Another welcome sign is the growing attention being paid to quality of life and to ecology...'[42] However, at the same time, the encyclical talks about the 'spreading of death caused by reckless tampering with the world's ecological balance...'[43]

This culture of death comes precisely from forgetting God:

> Moreover, once all reference to God has been removed, it is not surprising that the meaning of everything else becomes profoundly distorted. Nature itself, from being 'mater' (mother), is now reduced to being 'matter,' and is subjected to every kind of manipulation. This is the direction in which a certain technical and scientific way of thinking, prevalent in present–day culture, appears to be leading when it rejects the very idea that there is a truth of creation which must be acknowledged, or a plan of God for life which must be respected. Something similar happens when concern about the consequences of such a 'freedom without law' leads some people to the opposite position of a 'law without freedom,' as for example in ideologies which consider it unlawful to interfere in any way with nature, practically 'divinizing' it. Again, this is a misunderstanding of nature's dependence on the plan of the Creator. Thus it is clear that the loss of contact with God's wise design is the deepest root of modern man's confusion, both when this loss leads to a freedom without rules and when it leaves man in 'fear' of his freedom. By living 'as if God did not exist,' man not only loses sight of the mystery of God, but also of the mystery of the world and the

mystery of his own being.[44]

The document proposes the garden of the world as the object of ecology:

> As one called to till and look after the garden of the world (cf. Gn 2:15), man has a specific responsibility towards the environment in which he lives, towards the creation which God has put at the service of his personal dignity, of his life, not only for the present but also for future generations. It is the ecological question–ranging from the preservation of the natural habitats of the different species of animals and of other forms of life to 'human ecology' properly speaking — which finds in the Bible clear and strong ethical direction, leading to a solution which respects the great good of life, of every life.[45]

The text repeats that the lordship of man over the world is not absolute, but 'ministerial':

> Called to be fruitful and multiply, to subdue the earth and to exercise dominion over other lesser creatures (cf. Gn 1:28), man is ruler and lord not only over things but especially over himself, and in a certain sense, over the life which he has received and which he is able to transmit through procreation, carried out with love and respect for God's plan. Man's lordship however is not absolute, but ministerial: it is a real reflection of the unique and infinite lordship of God.[46]

The Pope teaches that a balanced ecological debate must consider the world both a 'home' and a 'resource':

> At the same time, *biblical anthropology* has considered man, created in God's image and likeness, as a creature who can transcend worldly reality by

virtue of his spirituality, and therefore, as a responsible custodian of the environment in which he has been placed to live. The Creator offers it to him as both a *home* and a *resource*.[47]

The Pope insists that it must be our relationship with God to condition our relationship with the environment:

> The consequence of this doctrine is quite clear: *it is the relationship man has with God that determines his relationship with his fellows and with his environment.* This is why Christian culture has always recognized the creatures that surround man as also gifts of God to be nurtured and safeguarded with a sense of gratitude to the Creator. *Benedictine and Franciscan spirituality* in particular has witnessed to this sort of kinship of man with his creaturely environment, fostering in him an attitude of respect for every reality of the surrounding world.[48]

If, on the other hand, solely pragmatism is pursued, the consequences are dramatic:

> In the *secularized modern age* we are seeing the emergence of a twofold temptation: a concept of knowledge no longer understood as wisdom and contemplation, but as power over nature, which is consequently regarded as an object to be conquered. The other temptation is the unbridled exploitation of resources under the urge of unlimited profit–seeking, according to the capitalistic mentality typical of modern societies.[49]

The solution to the problems is often found in the scientific and technological fields. In effect, technology which pollutes can also reduce pollution, and production which accumulates can also be distributed equitably, provided there is a prevailing ethic of respect for the life and dignity of human beings, and for the rights of present and future generations.[50]

The Pope pointed out that nature cannot be effectively defended if justification is claimed for acts which strike at the very heart of creation, which is human life. It is grossly inconsistent to oppose the destruction of the environment while allowing, in the name of comfort and convenience, the slaughter of the unborn and the procured death of the elderly and the infirm, and the carrying out, in the name of progress, of unacceptable interventions and forms of experimentation at the very beginning of human life. When the good of science or economic interests prevail over the good of the person, and ultimately of whole societies, environmental destruction is a sign of a real contempt for man.[51]

John Paul II noted that armed conflicts and the unrestrained race for economic growth are also among the factors that damage environmental equilibrium. The latter need to be moderated by measures carried out for the common good, and not just driven by earning capacity and personal profit. The ecological issue must be centred in a framework of harmonious human development, which considers people's cultural as well as scientific growth, and the formation of a mentality of respect and solidarity in more industrialized nations. Everyone, both scientists and politicians, must be involved. In order for the planet to be livable in the future and for everyone to have a place, the Pope has encouraged public authorities and all people of good will to examine their daily attitudes and decisions, which cannot be directed toward an endless and unbridled search for material goods that does not take into account the environment in which we live, but must rather aim to provide for the fundamental needs of present and future generations.[52] Such an approach to the environment basically demands the grace of conversion.[53]

The Pontifical Council for Culture indicated that the diffusion of scientific knowledge often leads man to envision himself in the immensity of the created world and to become enraptured with his own capacities and with the universe, without even thinking that God is its author. At the same time, a new awareness is emerging along with ecological development. This is not a novelty for the Church: the light of faith illuminates the meaning of creation and the relationship between man and nature. Saint Francis of Assisi and Saint Philip Neri are great witnesses to the respect for nature inscribed in the Christian view of the created world. This respect finds its origin in the fact that nature is not man's property; it belongs to God, its Creator, Who entrusted its stewardship to man (cf. Gn 1:28) so that he would respect it and find his legitimate sustenance in it.[54]

This idea of a conversion was adopted on later occasions by the Pope, indicating that human beings receive a mission of governing over creation in order to make its potential shine through. It is a delegation given by the Divine King at the very origin of creation when man and woman, who are the 'image of God' (Gn 1:27), receive the command to be fertile, multiply, fill the earth, subdue it and have dominion over the fish of the sea, the birds of the air and every living being that crawls on the earth (cf. Gn 1:28).[55] Saint Gregory of Nyssa, one of the three great Cappadocian Fathers, commented:

> the best Artificer made our nature as it were a formation fit for the exercise of royalty … it is the image of that Nature which rules over all … so the human nature also, as it was made to rule the rest,

> was, by its likeness to the King of all, made as it
> were a living image ... perfectly like to the beauty
> of its archetype in all that belongs to the dignity of
> royalty.[56]

John Paul II reaffirmed that the sovereignty of man is not absolute, but ministerial, and it is the true reflection of the unique and infinite sovereignty of God.[57] In biblical language, the 'naming' of the creatures (cf. Gn 2:19–20) is a sign of this mission of the knowledge and transformation of all created reality. It is the mission not of an absolute and unchallengeable master, but of a minister of the Kingdom of God, called to continue the Creator's work, a work of life and peace. His task, defined in the Book of Wisdom, is to 'govern the world with holiness and justice' (Ws 9:3). Unfortunately, if one takes a look at the regions of our planet, it is immediately evident that humanity has fallen short of this divine expectation. Especially in our time, man has uncontrollably devastated plains and wooded valleys, polluted the waters, disfigured the earth's habitat, rendered the air unbreathable, disrupted hydrogeological and atmospheric systems, desertified green areas, and carried out forms of unharnessed industrialization, humiliating our earthly home.[58]

Precisely in this context, it is urgent to encourage and sustain 'ecological conversion,' which in recent decades has made humanity more sensitive to the catastrophe toward which it has been heading. Man no longer seems the 'minister' of the Creator, but an autonomous despot. He is finally beginning to understand the importance of stopping before the abyss.

Therefore, it is not only a 'physical' ecology which is in play, attentive to safeguarding the habitats of various living beings, but also a 'human' ecology which renders the existence of these creatures more dignified, protecting the fundamental good of life in all of its manifestations and preparing an environment for future generations that more closely resembles the Creator's plan.[59] Conversion will stimulate a rediscovered harmony with nature and with oneself, and, just as the biblical Jubilee suggested (cf. Lv 25:8–13,23), men and women will once again walk in the garden of creation, seeking to act so that the goods of the earth are available to all and not only to a privileged minority.

In the midst of these marvels, we discover the voice of the Creator, which reaches us from the heavens and from the earth, by day and by night: a language whose sound is not heard, capable of surpassing all boundaries (cf. Ps 19:2–5).[60]

The Pope also treated particular situations in specific areas of the world, such as Oceania:

> Oceania is a part of the world of great natural beauty, and it has succeeded in preserving areas that remain unspoiled. The region still offers to indigenous peoples a place to live in harmony with nature and one another. Because creation was entrusted to human stewardship, the natural world is not just a resource to be exploited but also a reality to be respected and even reverenced as a gift and trust from God. It is the task of human beings to care for, preserve and cultivate the treasures of creation ... Yet the natural beauty of Oceania has not escaped the ravages of human exploitation ... The continued health of this and other oceans is crucial for the welfare of peoples not only in Oceania but in every part of the world. The natural resources of Oceania need to be protected against the harmful policies of some industrialized nations

and increasingly powerful transnational corporations which can lead to deforestation, despoliation
of the land, pollution of rivers by mining, over–
fishing of profitable species, or fouling the fishing–
grounds with industrial and nuclear waste. The
dumping of nuclear waste in the area constitutes an
added danger to the health of the indigenous population. Yet it is also important to recognize that
industry can bring great benefits when undertaken
with due respect for the rights and the culture of the
local population and for the integrity of the
environment.[61]

The *Common Declaration of John Paul II and the Ecumenical Patriarch His Holiness Bartholomew I*, signed in Venice on 11 June 2002, represents a moment of particular significance for the ecumenical aspect of the ecological question. The document reaffirms that respect for creation derives from respect for life and human dignity. Only if we recognize that the world was created by God can we discern an objective moral order within which to articulate a code of environmental conduct. In this perspective, Christians and all other believers have a specific task of proclaiming moral values and educating people to an ecological awareness, which is nothing other than taking on one's responsibility toward oneself, others, and creation.

The problem is not merely economic and technological; it is of a moral and spiritual order. A solution can be found at the economic and technological levels only if in the depth of our hearts there is as radical of a change as possible, which will lead us to change our lifestyle and our untenable models of consumption and production. A genuine conversion to Christ will allow us to change our ways of thinking and acting. The declaration first of all proposes the importance of recovering humility and recognizing the limits of our abilities, and, most importantly, the limits of our knowledge and of our judgment capacity. Second, we must frankly admit that humanity has a right to something greater than what we see around us. The value of prayer must be at the centre, imploring God the Creator to illuminate all people, wherever they may be, so that they feel the duty to respect and safeguard creation with care. The declaration claims, in a perspective of Christian optimism, that it is not too late. God's world has an incredible ability to heal. In the span of just one generation, with God's help and blessing, we could establish the correct orientation of the earth for our children's futures.[62]

The problem of the New Age ideology was also addressed by the Church in an important Vatican document, prepared by various departments. The most obvious implications of the New Age movement are:

> a process of conscious transformation and the development of ecology. The new vision which is the goal of conscious transformation has taken time to formulate, and its enactment is resisted by older forms of thought judged to be entrenched in the status quo. What has been successful is the generalisation of ecology as a fascination with nature and resacralisation of the earth, Mother Earth or *Gaia*, with the missionary zeal characteristic of Green politics. The Earth's executive agent is the human race as a whole ...[63]

The great danger of cosmocentrism, especially in relation to ecology, must also be identified as an integral part of this New Age ideology:

> Deep ecology's emphasis on bio–centrism denies the anthropological vision of the Bible, in which human beings are at the centre of the world, since

they are considered to be qualitatively superior to other natural forms. It is very prominent in legislation and education today, despite the fact that it underrates humanity in this way.. The same esoteric cultural matrix can be found in the ideological theory underlying population control policies and experiments in genetic engineering, which seem to express a dream human beings have of creating themselves afresh. How do people hope to do this? By deciphering the genetic code, altering the natural rules of sexuality, defying the limits of death.[64]

The response must be the integral Christian vision which promotes attention to the earth as God's creation. The question of respect for creation can be faced in a creative way by Catholics. Nonetheless, much of what has been proposed by the radical elements of the ecological movement is not readily reconcilable with the Catholic faith. In general, concern for the environment is an opportune sign of renewed attention for what God has provided, perhaps a necessary sign of Christian stewardship of creation, but so–called Deep Ecology is often based on pantheistic and at times gnostic errors.[65]

The *Compendium of the Social Doctrine of the Church* dedicates a chapter to environmental issues, recognizing the growing importance of these topics. The document calls everyone to an attitude of gratefulness and recognition toward creation. The text affirms, in fact, that the world reveals the mystery of God, Who created and sustains it. Rediscovering this profound meaning of nature helps us not only to rediscover God, but also to act responsibly toward the environment. The *Compendium* invites Christians to consider the environment with a positive outlook, avoiding a catastrophic vision and recognizing the presence of God in nature. We must look to the future with hope, the *Compendium* exhorts, 'sustained by the promise and the covenant that God continually renews.'[66] In the Old Testament we see how Israel lived the faith in an environment perceived as a gift from God. Furthermore, 'Nature, the work of God's creative action, is not a dangerous adversary.'[67] The *Compendium* recalls the beginning of the Book of Genesis, in which man is placed at the apex of all creatures and has the task entrusted to him by God of caring for all of creation. 'The relationship of man with the world is a constitutive part of his human identity. This relationship is in turn the result of another still deeper relationship between man and God.'[68] In the New Testament, Jesus uses natural elements in some of His miracles, reminding the disciples of the Father's Providence. Then, with His death and resurrection, 'Jesus inaugurates a new world in which everything is subjected to him and he creates anew those relationships of order and harmony that sin had destroyed.'[69]

Catholic tradition has recognized the progress achieved by science and technology in having amplified our possibilities for managing creation. Improving our living conditions in this world is in line with the will of God. The Church is not opposed to scientific progress, which is part of the human creativity given us by God. The *Compendium* adds nonetheless that: 'A central point of reference for every scientific and technological application is respect for men and women, which must also be accompanied by a necessary attitude of respect for other living creatures.'[70] Therefore, our use of the earth cannot be arbitrary, but must be ordered to a spirit of cooperation with God. The absence of this principle is what often lies at the root of actions which damage the environment. A reductionist understanding, which interprets the world in a 'mechanistic view,' together with the erroneous presupposition of limitless resources, leads to a vision of development in an exclusively material dimension, in which 'primacy is given to doing and having rather than to being'.[71] If we wish to avoid the error of reducing nature to purely utilitarian terms, in which it is considered merely something to exploit, we must also avoid moving to the opposite extreme of making it an absolute value. An ecocentric and biocentric view of the environment falls into the error of considering all creatures on the same level, ignoring the qualitative differences between human beings — founded on the dignity of the human person — and other creatures. The key to avoiding these errors is maintaining a Christian perspective. The *Compendium* explains that acting responsibly toward the environment is easier if we recall how God acted in creation. Christian culture considers creatures a gift from God, to be cared for and protected. Taking care of the environment is part of the responsibility of ensuring the common good, which also includes creation. This is a responsibility, the *Compendium* notes, which we also have toward future generations.

The *Compendium* further addresses the question of sharing the earth's resources. God created the goods of the earth to be used by all, the *Compendium* observes, and these goods 'must be shared equitably, in accordance with justice and charity.'[72] International cooperation on ecological issues is necessary, insofar as often the problems are of a global nature. The ecological issues are also often connected to poverty, for example where poor people do not have the possibility of addressing problems such as agricultural soil erosion due to a scarcity of economic and technological means. Many of these poor people live in

polluted city suburbs in make–shift shelters or in dilapi-
dated and dangerous house complexes. 'In such cases
hunger and poverty make it virtually impossible to avoid
an intense and excessive exploitation of the
environment.'[73] The response to these problems cannot be,
however, policies of demographic control which do not
respect the dignity of the human person. According to the
Compendium, demographic growth is 'fully compatible
with an integral and shared development.'[74] Development
must be integral, the text continues, and be directed
toward the true good of every person and of the entire
person. The principle of the universal destination of goods
must be applied to all natural resources, including water.
A vast number of people do not have sufficient access to
potable drinking water, which is often the cause of illness
and death. The *Compendium* finally offers some considerations on lifestyles that can help the developing world. At
the individual and social levels, the virtues of sobriety,
temperance and self–discipline are recommended. It is
necessary to escape from the logic of mere consumption
and to realize the ecological consequences of our decisions,
the document exhorts.

3.1.4 Benedict XVI

A key to interpreting the thought of Pope Benedict XVI is
offered to us in his homily at the Mass marking the beginning of his Pontificate:

> The external deserts in the world are growing,
> because the internal deserts have become so vast.
> Therefore the earth's treasures no longer serve to
> build God's garden for all to live in, but they have
> been made to serve the powers of exploitation and
> destruction. The Church as a whole and all her

> Pastors, like Christ, must set out to lead people out
> of the desert, towards the place of life, towards
> friendship with the Son of God, towards the One
> who gives us life, and life in abundance.[75]

The aesthetic dimension illustrates the contrast between the desert and the garden in this context: 'The beauty of nature reminds us that we have been appointed by God to "tend and care for" this "garden" which is the earth (cf. Gn 2: 8–17), and I see that you truly tend and take care of this beautiful garden of God, a true paradise. So, when people live in peace with God and one another, the earth truly resembles a "paradise." Unfortunately, sin ruins ever anew this divine project, causing division and introducing death into the world. Thus, humanity succumbs to the temptations of the Evil One and wages war against itself. Patches of "hell" are consequently also created in this marvellous "garden" which is the world.'[76]

The Pope then further explored the Eucharistic dimension of the environmental question, affirming that it is the Eucharist which heals the internal desert in human beings. The Holy Eucharist itself casts a powerful light upon human history and upon the entire universe. In this sacramental perspective we learn, day by day, that every ecclesial event has the nature of a sign through which God communicates Himself and speaks to us. In this way, the Eucharistic form of existence can truly favour an authentic mentality change concerning the way that we understand history and the world. The Liturgy itself educates us toward this, when, during the presentation of the gifts, the priest offers God a prayer of blessing and of petition over the bread and wine, 'fruit of the earth', 'fruit of the vine' and 'work of human hands.' With these words, in addition to involving all human work and activity in the offering to

God, the rite urges us to consider the earth God's creation, which produces for us what we need for our sustenance. It is not a neutral reality, merely matter to be used indifferently according to human instinct. Rather, it is part of God's good plan, in which we are all called to be sons and daughters in the only Son of God, Jesus Christ (cf. Ep 1:4–12). Justified concerns for the ecological conditions in which creation finds itself in many parts of the world find comfort in the perspective of Christian hope, which commits us to work responsibly to safeguard creation. In the relationship between the Eucharist and the world, in fact, we discover the unity of God's plan and we are led to grasp the profound relation between creation and the 'new creation,' which began with the resurrection of Christ, the New Adam. We participate in this already by virtue of our baptism (cf. Col 2:12f), and therefore in our Christian life, nurtured by the Eucharist, the perspective opens of a new world, a new heaven and a new earth, where the new Jerusalem comes down from heaven, from God, 'prepared as a bride adorned for her husband (Rev 21:2).[77]

Benedict XVI, following in the footsteps of the encyclical *Centesimus Annus* by Pope John Paul II, has indicated the importance of a human ecology, an ecology of peace, going beyond a merely physical ecology. In fact, there is a strong connection between the two ecologies, as illustrated by the daily worsening problem of *energy needs*.

> In recent years, new nations have entered enthusiastically into industrial production, thereby increasing their energy needs. This has led to an unprecedented race for available resources. Meanwhile, some parts of the planet remain backward and development is effectively blocked, partly because of the rise in energy prices. What will

happen to those peoples? What kind of development or non–development will be imposed on them
by the scarcity of energy supplies? What injustices
and conflicts will be provoked by the race for
energy sources? And what will be the reaction of
those who are excluded from this race? These are
questions that show how respect for nature is
closely linked to the need to establish, between
individuals and between nations, relationships that
are attentive to the dignity of the person and
capable of satisfying his or her authentic needs. The
destruction of the environment, its improper or
selfish use, and the violent hoarding of the earth's
resources cause grievances, conflicts and wars,
precisely because they are the consequences of an
inhumane concept of development. Indeed, if
development were limited to the technical–
economic aspect, obscuring the moral–religious
dimension, it would not be an integral human
development, but a one–sided distortion which
would end up by unleashing man's destructive
capacities.[78]

The Pope put the ecological question also within the
context of conscience and the natural law, in such a way
that it is not isolated also from a consideration of the
human person:

Today, we all see that man can destroy the foundations of his existence, his earth, hence, that we can
no longer simply do what we like or what seems
useful and promising at the time with this earth of
ours, with the reality entrusted to us. On the
contrary, we must respect the inner laws of creation, of this earth, we must learn these laws and
obey these laws if we wish to survive. Consequently, this obedience to the voice of the earth, of

being, is more important for our future happiness than the voices of the moment, the desires of the moment. In short, this is a first criterion to learn: that being itself, our earth, speaks to us and we must listen if we want to survive and to decipher this message of the earth. And if we must be obedient to the voice of the earth, this is even truer for the voice of human life. Not only must we care for the earth, we must respect the other, others: both the other as an individual person, as my neighbour, and others as communities who live in the world and have to live together.[79]

Pope Benedict XVI also addressed particular aspects of the environmental problem. On the twentieth anniversary of the adoption of the Montréal Protocol on substances that deplete the ozone layer, he spoke about the hole in the ozone layer: 'In the past two decades, thanks to exemplary collaboration in the international community between politics, science and economics, important results have been achieved with positive repercussions on the present and future generations. I hope that cooperation on everyone's part will be intensified in order to promote the common good and the development and safeguard of creation, strengthening the alliance between man and the environment, which must mirror the creative love of God from whom we come and to whom we are bound.'[80]

Benedict XVI has often expressed a heartfelt appeal to defend creation, in particular water, affirming that attention to climatic changes is very important. The Pope sent a greeting to the participants of a symposium entitled *The Arctic: Mirror of Life*. This symposium took place on the West coast of Greenland, under the sponsorship of His Holiness Bartholomew I, Ecumenical Patriarch of Constantinople. In the letter addressed to the Ecumenical Patriarch,

Benedict XVI encouraged prosperous nations to share their technology in favour of the environment with developing countries.[81]

The ecumenical aspect of the ecological question was particularly stressed: Pope Benedict XVI affirmed that it is more important than ever for all Christians to work together in arousing awareness of the issue, in order to show 'the intrinsic connection between development, human need and the stewardship of creation.'[82] The task of emphasizing an appropriate catechesis concerning creation, in order to recall the meaning and religious significance of its safeguarding, is intimately connected to the work of pastors and can have an important impact on the perception of the value itself of life and on an adequate solution of the consequent inevitable social problems. In particular, the Pope wrote: 'I see our common commitment as an example of that collaboration which Orthodox and Catholics must constantly seek, to respond to the call for a common witness. This implies that all Christians seriously cultivate the mental openness that is dictated by love and rooted in faith. Thus, they will be able together to offer to the world a credible witness of their sense of responsibility for the safeguarding of creation.'[83]

The Pope also reminded the younger generations of the importance of commitment to the environment:

> There is no doubt that one of the fields in which it seems urgent to take action is that of safeguarding creation. The future of the planet is entrusted to the new generations, in which there are evident signs of a development that has not always been able to protect the delicate balances of nature. Before it is too late, it is necessary to make courageous decisions that can recreate a strong alliance between

humankind and the earth. A decisive '*yes*' is needed to protect creation and also a strong commitment to invert those trends which risk leading to irreversibly degrading situations. I therefore appreciated the Italian Church's initiative to encourage sensitivity to the problems of safeguarding creation by establishing a National Day, which occurs precisely on 1 September. This year attention is focused above all on *water*, a very precious good which, if it is not shared fairly and peacefully, will unfortunately become a cause of harsh tensions and bitter conflicts.[84]

In this context, unfortunately it must be noted that many young people who are active in safeguarding the environment are inconsistent, insofar as they do not strive to promote human life, from conception to natural death.[85]

3.2 Episcopal teaching

3.2.1 German Bishops' Conference

Already by 1980, the German Bishops' Conference had published an important document with a particularly significant title: *The Future of Creation — the future of humanity*.[86] The origin of ecological problems was clearly expressed:

It is not acceptable for man to do everything he has the power to do. The greater this power, the greater his responsibility becomes. Together with the possibilities of promoting and improving life, the possibilities of damaging and destroying it also increase. Production and consumption growth do not necessarily indicate human development. Wherever the priority of spiritual goods over material goods is not respected, wherever the priority of

people over things is ignored, the internal and
external balance of peace — as well as the balance
of a just, worldwide social order — is put in
danger.[87]

Man sees himself as the focal point of earthly creation, and
the world as his home, which he organizes according to his
own utility and advantage. However thinking of doing
what is in his just interest, he runs the risk of using this
home in such a way that it collapses on him, leaving him
alone and helpless. Only through solidarity with the rest of
creation, only in responsible behaviour towards animals,
plants and the inanimate world, can he in the long run live
as master of creation, without becoming a slave to his
delusions of grandeur, a slave rejected by creation. For this
situation, the expression 'crisis of creation' could also be
used.[88]

In planning solutions to ecological problems, faulty
approaches should be avoided. The romantic dream of a
virgin world and a nature that cannot be touched ignores
history and underestimates the demands that the right to
life places on the current world population. Our predomi-
nant concern must be the possibility of life for all, in
particular for the poorest among us. Closing our eyes to the
fact that we can no longer continue to produce and
consume indefinitely as we now do, leads to the same
results: robbing each other, and robbing later generations
of the possibility of life. An attitude of panic toward a now
inevitable change in our lifestyles, and the refusal to
actively mould our future, attract that catastrophe which a
more rational approach could limit to the strictly necessary
and possible. Turning our backs on a situation which
seems unclear delays the inevitable decision, without
delaying its consequences; letting the responsibility fall to

others allows chance to dominate. In the end, it is not a question of planning for and granting a secure future to only a part of humanity — separating those who have the right to live tomorrow from those who do not — in order to avoid having to moderate our needs. Most importantly, environmental and energy questions cannot be resolved through the proposition of global–scale birth control.[89]

The God of the Covenant manifests Himself as the Creator of the heavens and the earth, the omniscient Father. If we see the world as a creation of God, it appears different to us and becomes new. It is the gift of a God Who loves and causes the preciousness of the world to grow. He Who gives is greater than the gift, and therefore the world becomes relative. The gift is at the same time also a task that we must become aware of, and therefore our responsibility increases.[90] Man and the world are not in a state of paradise. Since the relationship with God was disrupted by man's sin, a fracture runs through man's relationship with the rest of creation as well. Nonetheless, God preserves man's dual task concerning creation: to rule and to care for the world. However, man still has this painful experience: shaping the world works only in temporary ways, with great hardship and many risks. Accepting this burden can save man from the erroneous paths of idealistic utopias, from tired resignation and desperate violence.[91]

The Christian understanding of creation and the Christian relationship with it find their centre in Jesus Christ. He is the Word through Whom everything was created, the original divine model for creation. He does not remain an eternal idea beyond history, but rather He enters into it. In Him, God makes creation His own. Beginning with the Incarnation of the Son of God, the world becomes eternally part of God's life. Certainly the Son of God assumed our human nature to save us, but He has not cast it aside after having carried out the work of salvation. He wants to be and to remain what we are, His body and His soul are not only an instrument. By His hardships and suffering, He put Himself into our same conditions of life: working and suffering in this world. In His wonders and miracles, and primarily in His Resurrection, He has revealed to us the glory to which creation is called. Through His cross, frailty and finitude find their transformation and reconciliation; in His exaltation, the fulfilment of man and the world has already begun. The happy message of God's creation is maintained, recapitulated and surpassed through the happy message of Jesus Christ. There is no more profound reason, more radical measure or greater assurance for man's task concerning creation than Jesus Christ Himself.[92]

Later, in 1998, the German Bishops' Conference published another document, *Working for the future of creation*, also dealing with the environmental question. The document reveals the origins of the environmental crisis. One of the primary causes of human failure with regard to the environmental crisis is an understanding of science and technology which aims exclusively at the increase of human discretionary power over nature and does not include the protection of the environment in the technological–economic calculation. This leads to the complex problem of 'wasteful wealth in industrialized nations. The ecological crisis therefore finds another primary cause in consumption–oriented lifestyles, in which man's happiness lies in the continually growing, constant satisfaction of material needs. In their understanding of quality of life, people in modern society tend to look more toward material goods, possessions and consumption. Their philosophy of life is marked by the opinion according to which being is having, and having things is a form of fulfilled existence.[93]

Factors which weaken the perception of the consequences of human actions worsen the crisis. Concerning the ecological issue, this is valid by virtue of its structural nature: environmental damage often comes about as an indirect and unforeseen effect of technological–economic action, gradually becoming more severe. This makes predictive calculations of the dangers very complicated, and renders nearly impossible the identification of single causes according to the principle of causality. It should be added that many threats to the biosphere are caused by the reactive and synergetic effects of substances and procedures which, evaluated in themselves, are minimal and modest and therefore appear negligible. Deceived by appearance and not recognizing the significance of such small realities, a division between perception and reality occurs.[94] There is also a certain type of alarmist and emotionally–charged exaggeration regarding environmental dangers. Sometimes ecological issues are connected to a general apocalyptic attitude. The difficulty of forming an objective evaluation of potential ecological dangers has made social agreement on the necessary course of action very controversial. Such an accord has often been — and still is — blocked by ideological trench warfare.[95]

The German Bishops' Conference suggests that ecclesial action, which must be considered a stable response to the ecological crisis, demands reflection on the theological and ethical basis of a responsible relationship with creation. In this regard, the biblical starting point is the doctrine of creation, upon which the Christian approach toward the possibility of understanding and giving value to nature is founded. In this case, as will be explained shortly, various mutual incentives and points of agreement can be found between general environmental arguments and the theology of creation. In addition to the theological texts, reference must inevitably be made — even in the realm of Christian ethics — to the numerous current ethical–philosophical models in support of ethical–environmental claims. These models provide a series of concepts which can help convey the Church's input on both the scientific and social aspects of the environmental debate. At the same time, the translation of fundamental biblical themes into ethical–philosophical concepts provides a good opportunity to describe the specific nature of the Christian contribution. In a third phase, deeper ethical–theological reflections concerning the crucial idea of sustainable development must be presented. In this way, various links to the current debate on the environment relating to politics, economics and public opinion become apparent. Finally, the principle of sustainability is presented in relation to Christian social ethics and summarized in various general ethical criteria.[96]

3.2.2 Lombardy Bishops' Conference

The Lombardy Bishops' Conference published *The environmental question* in 1988.[97] The document first of all addresses the problem of ideologies:

> In any case, it is necessary to make a careful distinction between the general approval that must be accorded to many of the social and political appeals put forth from different environmental organizations, and the disapproval that must be declared against those who attempt to use the urgency of our ecological predicament as an appeal for what are complex global, civil and political schemes.[98]

The contrast between the land blessed by God and the soil cursed by original sin is also highlighted. It suggests how 'the gratification of bodily needs is not adequate for the fulfillment of human life. For man does not live by bread alone; to live, man has need for the Word.'[99] The bishops emphasize that: 'The so–called material wealth must be recognized as real wealth only on the condition that it becomes for man's conscience a symbol and token of hoped for spiritual wealth.'[100] Three ethical criteria are indicated as those which must guide man's interaction with the environment: respect and gratitude to God, moderation, and attentiveness to the quality of life.[101]

3.2.3 Portuguese Bishops' Conference

The document by the Portuguese Bishops' Conference, released on February 11, 1988 with the title *Pastoral note on the preservation of the environment*[102] primarily speaks about ecology as a cultural problem. Later the Conference dealt with the environmental question in the context of the common good. It affirmed that nature is part of the logic of giving and receiving: it is a gift which must be passed on to future generations, who have the right to receive an environment in better condition than before.[103]

3.2.4 Bishop of Talca, Chile

A fine example of a pastoral letter from a bishop is the one written by Mgr. Carlos Gonzalez, Bishop of Talca, Chile, and released on 4 October 1989. The issue was presented by relating the ecological problem with the culture of death. In opposition to this, the Gospel proposes a culture of life. There is therefore a need for a spirituality of life and celebration, which characterizes the path of the Beatitudes. Creation and sin are like the contrast between harmony

and disharmony. Human beings are God's grandest creation, and Jesus Christ shows us the face of God. The bishop
reveals that man is 'one' with nature, while at the same
time he is 'other' because of culture. Man is a natural being,
but also a 'cultural being.' It is necessary to rediscover his
initial solidarity with nature and with God. Nature and life
have rhythms which must be respected. Man is responsible
for the harmony of his world, and this responsibility
presupposes freedom and awareness. The violence of
nature and the violence of man, instead, lead to the
destruction of the world. In the dynamic between goodness and sin, the need for harmony is discovered. The fate
of the natural and cultural worlds is the responsibility of
the Church and of Christians.

Jesus Christ understands, reveals and creates the world,
and manifests God the Creator, Lord of all people. Jesus
accepts the human condition in contact with nature, and
institutes the Holy Eucharist, a sacrament of earth and toil,
which is also the pledge of a new world. Culture is an
integral reality, and synthesizes man's historicity. The
complexity of culture also leads to the need for evangeliza-
tion. The Gospel of Jesus, necessary for a more
authentically human life, draws us into the Kingdom with
pure hearts. The Kingdom grows slowly but strongly, and
the beatitudes, the road to the Kingdom, cast their
profound light on the major problems of our day. Modern
life, on the other hand, is marked by counter–beatitudes,
signs of the death which we live today. We must instead
follow the signs of Life, an authentic life in justice. Three
essential points are at the root of the solution to the environmental problem about which no one can remain
indifferent: justice, mercy and purity of heart.

3.2.5 United States Conference of Catholic Bishops

In November 1991, the United States Conference of Catholic Bishops issued an invitation to reflection and action on environment in the light of Catholic social teaching, entitled *Renewing the Earth*. The document aims to see ecological issues in relation to other areas of human and Christian concern:

> The web of life is one. Our mistreatment of the natural world diminishes our own dignity and sacredness, not only because we are destroying resources that future generations of humans need, but because we are engaging in actions that contradict what it means to be human. Our tradition calls us to protect the life and dignity of the human person, and it is increasingly clear that this task cannot be separated from the care and defence of all of creation.[104]

The American bishops' statement proposes six goals. First, in line with Papal teaching, to highlight the ethical dimensions of the environmental crisis. Then it links questions of ecology and poverty, environment and development. Third, the bishops stand with working men and women and poor and disadvantaged persons, whose lives are often affected by ecological abuse and compromises between environment and development. The document aims to promote a vision of a just and sustainable world community. Fifth, the bishops invite the Catholic community, and men and women of good will, to reflect more deeply on the religious dimensions of this topic. Finally the bishops desire to initiate a broader discussion on the potential contribution of the Church to environmental questions.

The document in a most perceptive way points out that nature is not, in Catholic teaching, merely a field to exploit at will or a museum piece to be preserved at all costs. We are not gods, but stewards of the earth.[105] One theological solution which the bishops propose is to view creation in a sacramental context as gift from God:

> The whole universe is God's dwelling. Earth, a very small, uniquely blessed corner of that universe, gifted with unique natural blessings, is humanity's home, and humans are never so much at home as when God dwells with them. In the beginning, the first man and woman walked with God in the cool of the day. Throughout history, people have continued to meet the Creator on mountaintops, in vast deserts, and alongside waterfalls and gently flowing springs... But as heirs and victims of the industrial revolution, students of science and the beneficiaries of technology, urban–dwellers and jet–commuters, twentieth–century Americans have also grown estranged from the natural scale and rhythms of life on earth. For many people, the environmental movement has reawakened appreciation of the truth that, through the created gifts of nature, men and women encounter their Creator. The Christian vision of a sacramental universe—a world that discloses the Creator's presence by visible and tangible signs—can contribute to making the earth a home for the human family once again.[106]

3.2.6 *National Conference of Brazilian Bishops*

In 1992, the National Conference of Brazilian Bishops drew up a document entitled *The Church and the ecological question*. The document outlines the ecological question in the realm of the challenge to the right to life, criticizing some developmental policies, and reevaluating the relationship between man and the environment. In the theological perspective of the Brazilian bishops based on Christian tradition, the Holy Spirit enables us to understand reality as energy and life. He is the creator and giver of life. He acts in all that moves, expands life, impassions the prophets, inspires poets, stirs up charismatic individuals and fills all of us with vigour, through which we continue to live and act. The Spirit fills the universe and renews the structure of the cosmos. With this idea of the Spirit, which penetrates the entire universe and is present in it, it is possible to rethink and offer a renewed dimension to the understanding of the relationship between God and creation. God is not distant, totally transcendent with respect to His creation. He abides in it, in the form of the life–giving Spirit. The Spirit made creation its home and remains always in relation with it. It is not a question of consecrating nature again. Nature, beginning with this, must also be perceived as the place of the presence of God Himself.[107] The document affirms that the current ecological crisis calls upon us, as a 'visit from God' and a moment of grace, to exercise solidarity toward creatures. Accepting this call, which reaches us from God through the wounded planet, leads us to the ethical foundations of our responsibility to the created world. This responsibility comes from our condition as creatures, demands a new relationship with nature, and asks of us a new spirituality and the exercise of ecological virtues.[108]

3.2.7 Federation of Asian Bishops' Conferences

In February 1993, a colloquium of the Federation of Asian Bishops' Conferences on faith and science was held, during the course of which a declaration was developed as an Asian response to the ecological crisis.[109] The document also deals with the current environmental problems in that region and the serious alteration of the cycles of the entire ecosystem, and more specifically the problems resulting from these alterations. First and foremost, it addresses the problems resulting from deforestation, especially soil erosion, which leads to flooding and drought. Second, it deals with the increase in the concentration of gases produced by fuel combustion, which leads to a general rise in the earth's temperature ('greenhouse effect'). Another problem was identified in the large–scale use of non–renewable energy resources, caused by consumerism, which could endanger the quality of life of future generations. It also cites the serious exploitation of marine resources deriving from pernicious fishing practices, ranging from mangrove removal to coral reef damage or the use of coastal waters as refuse dumps. There is also the dangerous accumulation of toxic materials in the atmosphere, rivers, lakes and oceans and even in soil, caused by inadequate or even non–existent treatment of the industrial effluents from chemical plants. Radiation released by leaks and low–quality systems in nuclear power plants expose people to the risk of tumours and genetic malformations. There is contamination of fruit and vegetables due to the indiscriminate use of pesticides, and long–term exhaustion of fertile lands due to the excessive use of chemical fertilizers. It highlights, furthermore, population shifts caused by major projects such as dam building, mine excavation, highway and superhighway construction and laying new railway tracks. Often, these problems are nothing but the symptoms of the ecological crisis being witnessed today, but the underlying causes include the serious destitution of a large part of the population, human greed which leads to unchecked consumerism, ignorance of environmental problems and a lack of recognition of the earth's mechanisms for defending life. Finally, the document points toward sustainable development and insists on the priority of 'being over having.'

The document highlights that the environmental question is a moral problem, and proposes various solutions, also in terms of good education. Among these solutions is the integration of science and faith. Science is giving important contributions to the understanding and resolution of ecological issues, but the faith dimension, which recognizes God as the Creator and believes in His active presence and contemplates the beauty of His creation, must also be emphasized. The idea of humankind as master of the universe, furthermore, should be substituted with the idea of a steward, responsible for the well–being of this world. Then the move must be made toward the integration of culture and science. Modern science and technology must not enter into conflict with traditional cultures that are in harmony with nature, but culture, science and technology must be reintegrated. The efforts of science and technology must be permeated by authentic human and cultural values, in order to put them at the service of humanity in the protection and stewardship of our fragile ecosystem. All of this must be accompanied by authentic human development. We need the 'development of the the whole person and of every human being.'[110] Authentic human development also refers to the correct relationship of individual persons with God, with nature and with society.

3.2.8 Council of European Bishops' Conferences

In Bad Honnef in Germany in May 2000, there was a consultation on the topic of creation spirituality and environmental politics. Responsibility toward creation must be considered an important element in the life of the Church. The Christian spirituality of creation is characterized by respect for the gifts of nature and by the willingness to share them with all people. On the basis of such spirituality, the Church can provide an essential contribution to the solution of environmental and developmental problems. The beauty and expressiveness of the Christian Liturgy, which develops throughout the entire liturgical year, has nature — transfigured by the mystery of Christ — as its essential source of symbols. The concrete activity of the Church for sustainable and just development is realized first of all in the formation of consciences through the proclamation of the Word and education. Model projects constitute a condition of the Church's credibility and an important encouragement of imitation, such as in the realms of renewable energy, ecological construction or an agricultural approach respectful of nature. Christian responsibility toward creation touches every area of the Church's action, from administration to pastoral care. Even lay ecclesial movements can show model commitment in this field.

3.2.9 Czech Bishops' Conference

The Czech Bishops Conference pointed out how communism destroyed the environment. Scientific communism, as an atheist ideology, considered man to be the master of history. It proclaimed that mankind had the key to its own fate and that it would build a paradise on Earth. In practice, though, promises of a future age of abundance in terms of material consumption contrasted with the low economic effectiveness of the Communist system. The concept of socialist industrialization with the goal first to catch up and overtake, and then to distribute to everybody according to their needs was simply a more vulgar and less successful variant of a discredited Enlightenment idea about permanent progress. Owing to the discrepancy between the aspirations and the possibilities, the growth of consumption in an ineffective system was often sought at the cost of the environment. The socialist economy did not even have mechanisms which would allow it to assess natural resources objectively. Consequently the ongoing devastation of water, air, trees, soil and nature in general was economically invisible. Also the effect on the health of the population only showed up with a delayed impact. Only after some years the level of consumption turned out to have been achieved at the cost of environmental damage; everything will be paid for by future generations.[111]

3.2.10 Australian Catholic Bishops' Conference

In 2002, the Australian Catholic Bishops' Conference published a document entitled *A New Earth — The Environmental Challenge*. It affirms above all that Christians believe that God created the universe and continually sustains it; God rejoices in all the creatures of the earth (Pro 8:30–31) and considers all of creation perfect (Gn 1:31). Saint Bonaventure saw the universe as 'a book reflecting, representing, describing its maker ...'[112] Since we are part of God's creation, we human beings are connected to all other creatures, to the natural world, and therefore to the entire universe. Unfortunately, human greed, violence and egoism have a destructive impact on people and the environment. Sin and its consequences in the world have everywhere damaged our relationship with God, ourselves, others and all of creation. Reconciliation is necessary. The life, death and resurrection of Jesus Christ, our Lord, bring salvation not only to the human race but also, in a different way, to the rest of creation. The document affirms that today it is an urgent duty of Christians to live reconciled with all of creation and to assume, in faith, the responsibility of caretakers of God's gifts: 'To achieve such reconciliation, we must examine our lives and acknowledge the ways in which we have harmed God's creation through our actions and our failure to act. We need to experience a conversion, or change of heart. God calls us to turn away from wrongdoing and to behave in new ways.'[113]

3.2.11 Bishops' Conference of England and Wales

In 2002, the Bishops' Conference of England and Wales issued a document entitled *The call of Creation: God's invitation and the human response*. The environmental crisis is especially complex since it involves not only many branches of scientific knowledge, but also politics and economics. The Church recognises and respects the autonomy of earthly affairs in all these disciplines. Its own task is to read the signs of the times and uncover the spiritual and moral issues that lie at the root of the challenges of our time.[114]

The environmental crisis has revealed the interdependence of all creation. Whatever we do, whatever choices we make, other people and the earth itself are affected. The symptoms of distress that have been outlined indicate that many human beings have lost an understanding of their true place in creation. By regarding the natural world merely as the setting in which we live, and by treating the gifts of creation solely for the satisfaction of our supposed needs as consumers, we have become alienated from the earth and from each other, and so also from God.[115]

Nature reveals God to us and allows us to experience God's presence. For example, people of faith have testified that nature's abundance and beauty reveals God's generosity and majesty, its healing, nourishing and life–giving properties reveal divine reconciling love. In the thirteenth century, St Thomas Aquinas argued that the diversity of the extraordinary array of creatures roaming the earth revealed the richness of the nature of God.

> And because one single creature was not enough, he produced many diverse creatures, so that what was wanting in one expression of the divine goodness might be supplied by another; for goodness, which in God is single and all together, in creatures is multiple and scattered. Hence the whole universe less incompletely than one alone shares and represents his goodness.[116]

It is also part of Christian faith to recognise that we are sinners. In our present context, this truth means that sin has distorted the human relationship with the natural world: we have disturbed the balance of nature in radical and violent ways. Sin damages our relationships with God and with one another, the relationships between social groups, and that between humanity and the earth.

The document is a reminder of the precious gifts of creation at each Eucharistic celebration.

> In the ancient prayer over the gifts of bread and wine we praise God our Creator, and remember that these material goods are given to us by God and are fashioned through the co–operation of Creator and creature: so our own daily living is to reflect our gratitude for the gifts that have been

given to us. Again, in the Eucharist we join in the self–giving, the sacrifice, of Christ himself, and in this sense the offering of our own lives — time, convenience, money — for the good of others can itself be Eucharistic, a 'sacrifice' for the good of others.[117]

3.2.12 Canadian Conference of Catholic Bishops

In 2003, the Social Affairs Commission of the Canadian Conference of Catholic Bishops published *The Christian Ecological Imperative*. The starting point is the beauty of the cosmos.

> The beauty and grandeur of nature touches each one of us. From panoramic vistas to the tiniest living form, nature is a constant source of wonder and awe. It is also a continuing revelation of the divine. Humans live within a vast community of life on earth. In the Jewish and Christian religious traditions, God is first described as the Creator who, as creation proceeded, 'saw that it was good.' God's love for all that exists was wondrously evident then, remains so now, and invites the active response of humankind.[118]

The document affirms that, while the glory of God is revealed in the world of nature, we human beings are destroying creation. From this perspective, the ecological crisis also appears to be a profoundly religious crisis. Therefore, a religious solution to the problem must be proposed. Over the course of history, the religious beliefs of every people have conditioned each one's relationship with the environment. Christians have developed a strong ecological sense like that of the saints. Pope John Paul II emphasized the need for an 'ecological conversion,'[119] and it is encouraging to see the numerous Christian traditions actively responding to the ecological crisis. Christian theological and liturgical tradition reaffirms the biblical message. Creation and the redeeming Incarnation of the Son of God are fully bound together. Through His Incarnation, Jesus Christ not only took up humanity, but He fully assumed it. He also embraced God's entire creation. Thus, all creatures, great and small, are consecrated in the life, death and resurrection of Christ. It is for this reason that the Church does not hesitate to bless and make abundant use of the earth's resources in liturgical celebrations and sacraments. Some people, however, have not been sufficiently aware of the problems posed by certain aspects of the Western capitalistic development model which have led to ecological collapse, to say nothing of the ecological disasters left in the wake of communist systems.[120]

3.3 Ecumenical aspects

3.3.1 Reformed perspectives

In 1985, a common declaration between the Catholic Church and the Evangelical Federation was promoted under the title: *Feeling responsibility for creation*. The document draws the following conclusions:

> The important thing, nonetheless, is that the churches and local communities communicate hope and clearly show how one cannot be held back by catastrophic fears in the assumption of one's responsibilities for God's creatures, but rather, trusting in the Divine Word, one must discover and develop the creative forces within man. In the Creed, Christians proclaim: 'I believe in God the Father Almighty, Creator of heaven and earth ...' All of Christianity believes in the Creator Who

created everything, 'heaven and earth,' and therefore also man with the earth as his 'natural habitat.'
With this, Christians recognize God's right over the
world and believe in the promise that the Creator is
also always the defender and Saviour. Those who
proclaim this statement of faith distinguish between
Creator and creature, and, in submission to God,
preserve the connection between the two realities.
Creation is subject to developing and being
destroyed ... for this reason we pray for the preservation of the world and hope in the redemption of
all creatures. Christians pray along with the psalm-
ist: 'The earth is the Lord's and all it holds, the
world and those who live there. For God founded it
on the seas, established it over the rivers' (Ps 24:1–
2).[121]

The Presbyterian communities in the United States
produced a paper in the 1990's which carried an interesting
reflection on tilling and keeping the earth (see Gn 2:15),
taken from the first two chapters of Genesis. Tilling
symbolizes everything humans do to draw sustenance
from nature.[122] It requires individuals to form communities
of cooperation and to establish systematic arrangements
(or economies) for satisfying their needs. Tilling includes
not only agriculture but manufacturing and exchanging,
all of which depend necessarily on taking and using the
gift of God's creation. Keeping the creation means tilling
with care, and so maintaining the capacity of the creation
to provide the sustenance for which the tilling is done.[123]
This means making sure that the world of nature may
flourish, with all its intricate, interacting systems upon
which life depends. However humans have failed to till
with care. The ecological crisis is the consequence of tilling
without keeping, together with the unfair distribution of

the fruits of tilling. The Creator's gifts for sustenance have not been taken carefully and shared equitably.

Adopting the biblical vision of tilling and keeping, one may say that sustainability is the capacity of those who till to keep the garden with sufficient care for tilling to continue. Moreover, since the garden is intrinsically good as God's creation, it is to be cherished not only for tilling but for its own sake. Sustainability is the capacity of the natural order and the socioeconomic order to thrive together. The steward is a manager, charged with responsibility for tilling and keeping for the sustenance of the household of humanity.[124]

3.3.2 Ecumenical Directory

In the Ecumenical Directory issued by the Pontifical Council for Promoting Christian Unity on 25 March 1993, the ecological question was addressed in an ecumenical light.

> There is an intrinsic connection between development, human need and the stewardship of creation. For experience has taught us that development in response to human needs cannot misuse or overuse natural resources without serious consequences. The responsibility for the care of creation, which in itself has a particular dignity, is given by the Creator himself to all people, in so far as they are to be stewards of creation. Catholics are encouraged to enter, at various levels, into joint initiatives aimed at study and action on issues that threaten the dignity of creation and endanger the whole human race. Other topics for such study and action could include, for example, certain forms of uncontrolled rapid industrialization and technology that cause pollution of the natural environment with serious

consequences to the ecological balance, such as destruction of forests, nuclear testing and the irrational use or misuse of both renewable and unrenewable natural resources.[125]

3.3.3 *Orthodox View*

The whole of the universe worships and offers gifts to its Creator. In the very shape of the churches and the placing of the icons, mosaics or frescoes within them, Orthodoxy indicates a microcosm of the universe, which clarifies the role both of humanity and of the rest of creation in relation to God. For it is an expression not just of what is on earth today, but of what exists in heaven and what is to come — the eschatological promise and redemptive transformation of all creation through the salvation wrought by Jesus Christ (see Romans 8).

In the view of the Orthodox Church, ecology is considered in its proper ecclesial perspective with a cosmic ecclesiology. It begins with the biblical notion of benediction (in Hebrew *berekh* – Gn 27:25–30); the idea is then linked to sacramentality and liturgical practice. Then ecology is set in the context of humanity's priestly vocation. The dominion of man over nature is understood as the first liturgical place of service. Thus, ecology is seen in the context of the benediction rites of the Orthodox Church. The blessing of water for baptism acquires a truly cosmic and redemptive significance. God created the world and blessed it and gave it to us as our food and life, as the means of communion with him. The blessing of water signifies the return or redemption of matter to this initial and essential meaning. By accepting the baptism of John, Christ sanctified the water — made it the water of purification and reconciliation with God. It was then, as

Christ was coming out of the water, that the Epiphany —
the new and redemptive manifestation of God — took
place, and the Spirit of God, who at the beginning of
creation moved upon the face of the waters, recreated
water — that is, the world — again into what He made it
at the beginning. In the blessing of the water on the Feast
of the Epiphany, creation must be restored and cleansed.
One prayer for the blessing of the water runs like this:

> Confer upon it the grace of redemption, the blessing
> of the Jordan. Make it a source of incorruption, a gift
> of sanctification, a remission of sins, a protection
> against disease, a destruction to demons, inacces-
> sible to the adverse powers and filled with angelic
> strength: that all who draw from it and partake of it
> may have it for the cleansing of their soul and body,
> for the healing of their passions, for the sanctifica-
> tion of their dwellings and for every purpose that is
> expedient.

The Church cannot pray for the water to be redemptive if
it is polluted. Ecology is considered in the context of
oikumene: the object of the divine economy is the *oikumene*
— all of the inhabited earth. All of creation is recapitulated
in Christ. Thus, the house is transformed into a temple. The
practice of ecology must be an ecclesial event, considering
the model of Noah's ark a prefiguration of the Church as
seen in 1 Peter 3:20–21.[126]

I. Zizioulas also proposes ecology in the liturgical
context, especially centred upon the Eucharist.[127] To bless
is to give thanks. In and through thanksgiving, we
acknowledge the true nature of things we receive from
God and thus enable them to attain the fullness God
intended for them. We bless and sanctify things when we
offer them to God in a eucharistic movement of our whole

being. And as we stand before the cosmos, before the matter given to us by God, this eucharistic movement becomes all–embracing. In general, it can be affirmed that the Orthodox tradition emphasizes the priestly vocation of humanity, and speaks the language of gift and benediction rather than of dominion and stewardship. The primordial relationship of Adam to both God and Creation is restored in the Eucharist and we have a foretaste of the eschatological state of creation.

The vocation of humanity, as shown in liturgical theology, is not to dominate and exploit nature, but to transfigure and hallow it. In a variety of ways — through the cultivation of the earth, through craftsmanship, through the writing of books and the painting of icons — humanity gives material things a voice and renders the creation articulate in praise of God.

The Orthodox Church shares the sensibility and commitment of those who are concerned with the increasing damage to the natural environment due to human abuses which the Church calls sin, and for this reason it invites all human beings to repentance. There is a tendency to seek a renewal of ethics, while the Orthodox Church retains that the solution must be found in the liturgical, Eucharistic and ascetic *ethos* of the Orthodox tradition. The Orthodox Church must not be identified with a movement, political party or organization, nor must its position be considered from an ideological or philosophical point of view, nor that of methods and programs applied for a solution to the ecological problem. The Orthodox Church constitutes a presence and a witness to a new way of existence that follows its specific theological approach to the relationship of human beings with God, with others and with nature. The Orthodox affirm that humanity needs a simpler way of life, a renewed asceticism, for the good of creation.[128] The Orthodox celebrate 1st September as a day of reflection on creation.[129] Incidentally, the Anglicans celebrate the so–called Harvest Festival in autumn. Furthermore, the holiday of Thanksgiving, celebrated in the United States on the last Thursday in November, is of clearly Christian inspiration, as the first pilgrims' thanks to God for the fruits of the earth.

3.3.4 World Council of Churches

An important document by the World Council of Churches was developed in Seoul, South Korea, in 1990.[130] It deals with ecology more from the point of view of social justice than from a vision of creation and redemption. The documents following it dealt with the question in terms of safeguarding creation, the spirituality of creation and responsibility for the environment. For the Council, undertakings for safeguarding creation must be carried out in correspondence with the critiques of the dominant economic development model. According to them, the liberal idea of limitless growth conflicts, structurally speaking, with the limited, non–renewable environmental resources. The economic interests of industrialized nations amount to direct attempts against creation. Therefore, to safeguard creation and make it more 'fruitful' for future generations as well, it is necessary to work toward an international economic order founded on justice, fair resource distribution, sharing and solidarity.

Notes

1   Pope St Clement I, *First Letter to the Corinthians*, 20.
2   Cf. Pope Pius XII, *Ad agrorum Cultores ob Conventum Confoe-*

derationis nationalis Italicae Romae coadunatos (15 November 1946), in *AAS* 38 (1946), p. 432. Cf. also Pliny, *Naturalis Historia*, Book III, 5, n.41.

3   Cf. Pope Pius XII, *Ad agrorum Cultores ob Conventum Confoederationis nationalis Italicae Romae coadunatos* in *AAS* 38 (1946), p. 432.

4   *Ibid.*, 1 in *AAS* 38 (1946), p.434: 'Colpito successivamente da tutti i flagelli, diluvi, cataclismi tellurici, miasmi pestilenziali, guerre devastatrici, il suolo in alcune parti deserto, sterile, malsano, ... si è rifiutato di elargire spontaneamente all'uomo i suoi tesori. La terra è la grande ferita, la grande malata. Chinato su di lei, non come lo schiavo sulla gleba, ma come il clinico sul letto del paziente, il coltivatore le prodiga le sue cure con amore.'

5   Cf. *ibid.*

6   Pope Pius XII, *Easter Radio Message* (10 April 1955) in *AAS* 47 (1955), pp. 284–285: 'Non con ansia e trepidazione abbiamo osservato i recenti progressi che, dopo alcuni impianti fissi (di produzione di energia elettro–nucleare), hanno condotto a buon termine il primo tentativo di muovere una nave con energia ricavata da trasmutazioni nucleari, mettendo finalmente queste forze a servizio, non a distruzione dell'uomo... Con pari fiducia ed attesa seguiamo quelle molte ricerche le quali, volte a studiare gli effetti che i numerosi tipi di radiazione ora disponibili hanno sui vegetali, sulla loro possibilità di conservazione ... Tuttavia a riguardo di ciò che la ricerca può fare nel dominio geloso della vita, dobbiamo ancora una volta ammonire dei pericoli, che la genetica prevede come possibili, quando il mistero, che giace in fondo a ciò che è vivo, viene manomesso da incauti interventi o da un violento mutamento dell'*habitat*, per esempio a causa di agenti, come un'accresciuta radioattività nei confronti di un'ancora ignota soglia di sicurezza biologica.'

7   Vatican II, *Gaudium et spes*, 34.1.

8   Pope Paul VI, *General Audience* (31 March1971) in *IP* 9 (1971), p. 242: 'Oggi ci si occupa di ecologia, cioè di purificazione dell'ambiente fisico dove si svolge la vita dell'uomo: perché non ci preoccuperemo anche d'un'ecologia morale dove l'uomo vive da uomo e da figlio di Dio.'

9   Pope Paul VI, *Octogesima Adveniens*, 21 in *IP* 9 (1971), pp. 1182–1183.

10 Pope Paul VI, *Message to the Stockholm Conference on Human Environment* (1 June 1972) in *IP* 10 (1972), pp. 606–610.

11 Pope Paul VI, *Discourse to the participants in the Study Week promoted by the Pontifical Academy of Sciences on the theme 'Natural Products and Plant Protection'* (23 October 1976) in *PA*, p.210. See also Idem, *Discourse to the Pontifical Academy of Sciences* (19 April 1975) in *AAS* 67 (1975), p. 268.

12 Cf. Pope Paul VI, *Message on the occasion of the 5th World Day for the Environment* (5 June 1977) in *IP* 15 (1977), pp. 561–562.

13 Pope John Paul II, *Redemptor Hominis* (1979), 15.

14 *Ibid.*, 16.

15 Pope John Paul II, *Commemoration of the centenary birth of Albert Einstein* (10 November 1979) in *PA*, p. 240.

16 Pope John Paul II, Apostolic Letter *Inter sanctos* (29 November 1979) in *AAS* 71 (1979), pp. 1509–1510: 'meritatamente annoverare tra i santi e illustri uomini che ebbero una singolare venerazione per la natura, quale magnifico dono fatto da Dio all'umanità... Egli infatti, ebbe una particolare percezione di tutte le opere del Creatore, e quasi superbamente ispirato compose quel bellissimo Cantico delle creature, attraverso le quali, in particolare fratello sole, sorella luna e le stelle, diede all'altissimo, onnipotente e buon Signore la debita lode, onore, gloria e ogni benedizione.' See also M. De Marzi, *San Francesco d'Assisi e l'ecologia* (Roma: Pontificio Istituto Antoniano, 1981).

17 Pope John Paul II, *Discourse to the participants in the Study Week on 'Energy and Humanity'* (14 November 1980) in *PA*,

pp. 245–247.

18 Pope John Paul II, *Address to the members of the Pontifical Academy of Sciences* (23 October 1982) in *IG* 5/3 (1982), p. 892. Cf. *CCC* 2418.

19 Pope John Paul II, *Address to the participants in the Congress on 'Environment and Health'* (24 March 1997), 5.

20 Pope John Paul II, *Address to the people of Assisi* (12 March 1982) in *IG* 5/1 (1982), pp.852–853: 'San Francesco sta dinanzi a noi anche come esempio di inalterabile mitezza e di sincero amore nei confronti degli esseri irragionevoli, che fanno parte del creato. In lui riecheggia quell'armonia che è illustrata con parole suggestive dalle prime pagine della Bibbia: "Dio pose l'uomo nel giardino di Eden, perché lo coltivasse e lo custodisse" (Gn 2:15), e "condusse" gli animali "all'uomo, per vedere come li avrebbe chiamati" (Gn 2:19)'; on the Christian notion of dominion, see Chapter 4, paragraph 4.3.3 below.

21 Cf. Pope John Paul II, *Address to the people of Assisi* (12 March 1982) in *IG* 5/1 (1982), pp. 852–853: 'come padroni e custodi intelligenti e nobili,' 'non come sfruttatori e distruttori senza alcun riguardo'; and also: Idem, *Redemptor Hominis*, 15.

22 Pope John Paul II, *Address to the people of Assisi* (12 March 1982) in *IG* 5/1 (1982), pp. 852–853: 'L'educazione al rispetto per gli animali ed, in genere, per l'armonia del creato ha, del resto, un benefico effetto sull'essere umano come tale, contribuendo a sviluppare in lui sentimenti di equilibrio, di moderazione, di nobiltà ed abituandolo a risalire "dalla grandezza e bellezza delle creature" alla trascendente bellezza e grandezza del loro Autore (cf. Sap 13:5).'

23 Pope John Paul II, *Address at the Liturgy of the Word in Zamosc* (12 June 1999), 3.

24 Pope John Paul II, *Homily at the Mass in the Cathedral of St. Stephen in Vienna* (12 September 1983).

25 Pope John Paul II, Encyclical Letter *Dominum et vivificantem*, 50.

26 Pope John Paul II, *Address to the participants in the Plenary Assembly of the Pontifical Academy of Sciences on the 50th Anniversary of Foundation* (28 October 1986), 8., in *PA*, p. 286.

27 Pope John Paul II, *Address to the participants in the Study Week organized by the Pontifical Academy of Sciences* (6 November 1987) in *IG* 10/3 (1987), p. 1018: 'Science is a human work and must be directed solely to the good of humanity. Technology, as the transfer of science to practical applications, must seek the good of humanity and never work against it. Therefore science and technology must be governed by ethical and moral principles. Theory aimed only at profit has produced in the last century a technology that has not always respected the environment, that has led to situations causing great concern by reason of the irreversible damage done, both locally and worldwide.'

28 Pope John Paul II, *Sollecitudo rei socialis* (30 December 1987), 34.

29 The complete document is found in Appendix 1. Here we offer a brief summary.

30 Pope John Paul II, Encyclical Letter *Centesimus Annus* (1 May 1991), 37.

31 Pope John Paul II, *Address to the participants in the international award for the environment: St. Francis, 'The Canticle of Creatures'* (25 October 1991) [translation mine]: 'I beni della terra, che nel piano divino debbono essere patrimonio comune, rischiano talora di diventare monopolio soltanto di pochi. Essi vengono utilizzati ad esclusivo beneficio di alcuni, che non di rado li manomettono e, talora, li distruggono, arrecando così un danno all'intera umanità. Occorre frenare la corsa all'uso egoistico dei beni della terra. Bisogna impedirne la distruzione e l'alterazione perché subiamo tutti le conseguenze negative di scelte ecologiche sconsiderate.'

32 Cf. *CCC* 299.

33 Cf. *CCC* 358.

34 Cf. *CCC* 373.

35   Cf. *CCC* 377.

36   *CCC* 342.

37   *CCC* 2415.

38   *CCC* 2416–2418.

39   Pope John Paul II, *Veritatis splendor* (6 August 1993), 38.

40   *Ibid.*, 39.

41   Pope John Paul II, *Address to the participants in the Workshop on Chemical Hazards in Developing Countries* (22 October 1993).

42   Pope John Paul II, *Evangelium vitae* (25 March 1995), 27.

43   *Ibid.*, 10.

44   *Ibid.*, 22.3.

45   *Ibid.*, 42.3.

46   *Ibid.*, 52.

47   Pope John Paul II, *Address to the participants in the Congress on 'Environment and Health'* (24 March 1997), 3.3.

48   *Ibid.*, 4.1.

49   *Ibid.*, 4.2.

50   Cf. *ibid.*, 5.4.

51   See Pope John Paul II, *Address at the Liturgy of the Word in Zamosc* (12 June 1999), 3.

52   Pope John Paul II, *Address to the members of the Pontifical Academy of Sciences* (12 March 1999), 2–3: 'In today's world, more and more people condemn the increasing harm caused by modern civilization to persons, living conditions, climate and agriculture. Certainly, there are elements linked to nature and its proper autonomy, against which it is difficult, if not impossible, to struggle. Nevertheless, it is possible to say that human behaviour is sometimes the cause of serious ecological imbalance, with particularly harmful and disastrous consequences in different countries and throughout the world. It suffices to mention armed conflict, the unbridled race for economic growth, inordinate use of resources, pollution of the atmosphere and water. Man has the respon-

sibility of limiting the risks to creation by paying particular attention to the natural environment, by suitable intervention and protection systems considered especially from the viewpoint of the common good and not only of viability or private profit. The sustainable development of peoples calls on everyone to place themselves "at the service of all, to help them to grasp this serious problem in all its dimensions, and to convince them that solidarity in action ... is a matter of urgency" (*Popolorum progressio*, n. 1).'

53 *Ibid.*, 7: 'People sometimes have the impression that their individual decisions are without influence at the level of a country, the planet or the cosmos. This could give rise to a certain indifference due to the irresponsible behaviour of some individuals. However, we must remember that the Creator placed man in creation, commanding him to administer it for the good of all, making use of his intelligence and reason. From this, we can be assured that the slightest good act of a person has a mysterious impact on social transformation and shares in the growth of all. On the basis of the covenant with the Creator, to which man is called to turn continually, everyone is invited to a profound personal conversion in their relationship with others and with nature. This will enable a collective conversion to take place and lead to a life in harmony with creation. Prophetic actions, however slight, are an opportunity for a great number of people to ask themselves questions and to commit themselves to new paths. Consequently, it is necessary to ensure that everyone, particularly young people who desire a better social life in the midst of creation, is educated in human and moral values; it is also necessary to develop every person's social sense and attentiveness to others, so that all may realize what is at stake in their daily attitudes for the future of their country and the world.'

54 Cf. Pontifical Council for Culture, *Towards a Pastoral approach to Culture* (1999), 11.

55 Pope John Paul II, *General Audience* (17 January 2001), 2.

56   St Gregory of Nyssa, *On the Making of Man*, IV, in P. Schaff –
     H. Wace (ed.), *Nicene and Post–Nicene Fathers, Gregory of
     Nyssa: Dogmatic Treatises, etc.*, vol. 5, (Peabody: Hendrickson
     Publishers, ²1995) pp. 390–391.

57   Pope John Paul II, *General Audience* (17 January 2001), 3. Cf.
     John Paul II, *Evangelium vitae*, 52.

58   Cf. Pope John Paul II, *General Audience* (17 January 2001), 3.

59   Cf. *ibid.*, 4.

60   Cf. *ibid.*, 5.

61   Pope John Paul II, Apostolic Exhortation *Ecclesia in Oceania*
     (22 November 2001), 31.

62   Cf. Pope John Paul II, *Common Declaration of John Paul II and
     the Ecumenical Patriarch His Holiness, Bartholomew I* (11 June
     2002), as presented in Appendix 2 below.

63   Pontifical Council for Culture and Pontifical Council for
     Interreligious Dialogue, *Jesus Christ the Bearer of the Water of
     Life. A Christian reflection on the 'New Age'* (30 January 2003),
     2.3.1.

64   *Ibid.*, 2.3.4.1.

65   Cf. *ibid.*, 6.2.

66   Pontifical Council for Justice and Peace, *Compendium of the
     Social Doctrine of the Church* (2004), 451.

67   *Ibid.*

68   *Ibid.*, 452.

69   *Ibid.*, 454.

70   *Ibid.*, 459.

71   *Ibid.*, 462.

72   *Ibid.*, 481.

73   *Ibid.*, 482.

74   *Ibid.*, 483.

75   Pope Benedict XVI, *Homily at the Mass for the Beginning of the
     Petrine Ministry* (24 April 2005).

76 Pope Benedict XVI, *Angelus* (22 July 2007).

77 Pope Benedict XVI, Post–Synodal Apostolic Exhortation *Sacramentum Caritatis* (22 February 2007), 92.

78 Pope Benedict XVI, *Message for the Celebration of the World Day of Peace* (1 January 2007). Then, in his letter to Professor Mary Ann Glendon, President of the Pontifical Academy of Social Sciences, on the occasion of the 13th Plenary Session (28 April 2007), he indicated that particular attention must be given to the fact that the poorest nations are those which seem destined to pay the highest price for ecological deterioration.

79 Pope Benedict XVI, *Meeting with the clergy of the dioceses of Belluno–Feltre and Treviso* (24 July 2007).

80 Pope Benedict XVI, *Angelus* (16 September 2007). The Montréal Protocol, carrying out what was determined by the Vienna Convention of 1985, established objectives and means for reducing the production and use of substances considered dangerous to the stratospheric ozone layer. The Protocol established the deadlines by which the signing parties must work to contain production and consumption levels of dangerous substances, and regulate commercial trade, communication of monitoring data, research activities, information exchange and technical assistance.

81 Cf. Pope Benedict XVI, *Letter to the Ecumenical Patriarch of Constantinople on the Occasion of the Seventh Symposium of the Religion, Science and the Environment Movement* (1 September 2007).

82 Cf. Pope Benedict XVI, *Letter to the Ecumenical Patriarch, His Holiness Bartholomew I, Ecumenical Patriarch, on the Occasion of the Sixth Symposium on 'Religion, Science and the Environment' Focusing on the Amazon River* (6 July 2006). Cf. Pontifical Council for Promoting Christian Unity, *Directory for the Application of the Principles and Norms of Ecumenism* (1994), 215.

83 Pope Benedict XVI, *Letter to the Ecumenical Patriarch, His*

Holiness Bartholomew I, Ecumenical Patriarch, on the Occasion
of the Sixth Symposium on 'Religion, Science and the Environment' Focusing on the Amazon River (6 July 2006).

84   Pope Benedict XVI, *Homily on the Occasion of the* Agorà *of
Italian Youth* (2 September 2007).

85   Cf. Pope Benedict XVI, *Address of His Holiness Benedict XVI to
H.E. Mr. Noel Fahey, New Ambassador of Ireland to the Holy See*
(15 September 2007): 'How disturbing it is that not infre-
quently the very social and political groups that, admirably,
are most attuned to the awe of God's creation pay scant
attention to the marvel of life in the womb. Let us hope that,
especially among young people, emerging interest in the
environment will deepen their understanding of the proper
order and magnificence of God's creation of which man and
woman stand at the centre and summit.'

86   German Bishops' Conference, *The Future of Creation — the
future of humanity* (1980).

87   *Ibid.,* I, 1.

88   *Ibid.*

89   Cf. *ibid.,* I, 3.

90   Cf. *ibid.,* II, 1.

91   Cf. *ibid.,* II, 4.

92   Cf. *ibid.,* II, 7.

93   German Bishops' Conference, *Safeguarding the future of creation* (1998), 29, 35.

94   Cf. *ibid.,* 36.

95   Cf. *ibid.,* 37.

96   Cf. *ibid.,* 56–58.

97   Lombardy Bishops' Conference, *La questione ambientale* [*The
environmental question*], (Milan: Centro Ambrosiano, 1988).

98   *Ibid.,* pp.13–14. Translation from D. Christiansen and W.
Grazier (eds.), *And God Saw That It Was Good: Catholic
Theology and the Environment* (Washington, DC: US Catholic

Conference, 1996), p. 299. 'In ogni caso occorre operare un'attenta distinzione tra il consenso che può e deve essere accordato a molte delle singole istanze sociali e politiche avanzate dai diversi movimenti ambientalisti, e il dissenso che invece dev'essere dichiarato nei confronti di chi pretenda trasformare una presunta istanza ecologica nell'equivalente di un progetto civile e politico complessivo e globale.'

99 Lombardy Bishops' Conference, *La questione ambientale*, p.18. Translation from D. Christiansen and W. Grazier (eds.), *And God Saw That It Was Good*, p. 301. 'La semplice soddisfazione dei bisogni non basta a realizzare la vita dell'uomo; di pane soltanto l'uomo non vive; per vivere egli ha bisogno di una parola.' Cf. also Mt 4:4; Dt 8:3.

100 Lombardy Bishops' Conference, *La questione ambientale*, p.20 Translation from D. Christiansen and W. Grazier (eds.), *And God Saw That It Was Good*, p. 302. 'I beni cosiddetti "materiali" possono e debbono essere riconosciuti quali veri beni soltanto a condizione che essi diventino per la coscienza dell'uomo segno e pegno dei beni sperati.'

101 Lombardy Bishops' Conference, *La questione ambientale*, pp. 24–26.

102 Original Portuguese title: *Nota pastoral sobre a preservação do meio ambiente.*

103 Portuguese Bishops' Conference, Pastoral Letter *Responsabilidade solidária pelo bem comum* [*Responsibility in solidarity for the common good*] (2003), 20: 'O meio ambiente é um dos bens comuns essenciais à vida da humanidade, é uma condição absolutamente necessária para a vida social ... O ambiente situa–se na lógica da recepção: é um empréstimo que cada geração recebe e deve transmitir à geração seguinte. Daí a enorme responsabilidade quanto ao uso e usufruto dos bens comuns ambientais em cada presente histórico. As gerações futuras têm o direito de receber o ambiente em melhores condições do que as

situações em que as gerações anteriores o viveram.'

104 United States Conference of Catholic Bishops, *Renewing the Earth* (1991), I, A.

105 See *ibid.*, I, C.

106 See *ibid.*, III, A.

107 Cf. National Conference of Brazilian Bishops, *The Church and the ecological question* (1992), 53–54.

108 Cf. *ibid.*, 61.

109 Federation of Asian Bishops' Conferences, 'Love for creation: An Asian Response to the Ecological Crisis' in *Catholic International* 4/6 (June 1993), pp. 269–272. See also Catholic Bishops' Conference of the Philippines, Pastoral Letter on Ecology, *What is Happening to our Beautiful Land?* (29 January 1988).

110 Pope John Paul II, *Sollicitudo rei socialis*, 32.

111 Czech Bishops' Conference, *Peace and good* (17 November 2000), 46.

112 Australian Catholic Bishops' Conference, *A New Earth — The Environmental Challenge* (2002). Cf. St Bonaventure, *Breviloquium* 2, 12. Pope John Paul II used the same image, sustaining that 'creation ... is almost another sacred book whose letters are represented by the multitude of created things present in the universe' (Cf. Pope John Paul II, *General Audience* [30 January 2002], 6).

113 *Ibid.*.

114 See Bishops' Conference of England and Wales, *The call of Creation: God's invitation and the human response*, I. See Vatican II, *Gaudium et Spes*, 36.

115 See Bishops' Conference of England and Wales, *The call of Creation: God's invitation and the human response*, III.

116 St Thomas Aquinas, *Summa Theologiae*, I, q.47, a.1. See Bishops' Conference of England and Wales, *The call of Creation: God's invitation and the human response*, IV.

117 See Bishops' Conference of England and Wales, *The call of*

Creation: God's invitation and the human response, VI.

[118] Canadian Conference of Catholic Bishops, *The Christian Ecological Imperative* (2003), 1.

[119] Sr M. Keenan, *From Stockholm to Johannesburg: An Historical Overview of the Concern of the Holy See for the Environment, 1972–2002* (Vatican City: Pontifical Council for Justice and Peace, 2002), p. 75.

[120] Canadian Conference of Catholic Bishops, *The Christian Ecological Imperative* (2003), 4, 7.

[121] Common Declaration of the Catholic and Evangelical Churches in Germany, *Feeling responsibility for creation* (14 May 1985), 99 as found in an Italian version in *Enchiridion Oecumenicum Supplementi — Dialoghi ecumenici (1984–1989)*, (Bologna: EDB, 1989), 181: 'L'importante comunque è che le chiese e le comunità locali comunichino speranza e mostrino chiaramente come non ci si può lasciar frenare da paure apocalittiche nell'assunzione delle proprie responsabilità per le creature di Dio, ma piuttosto, nella fiducia della parola divina, si lascino scoprire e crescere quelle forze creatrici che sono nell'uomo. Nel Credo i cristiani proclamano: "Credo in Dio Padre onnipotente, Creatore del cielo e della terra". L'intera cristianità crede nel Creatore che ha creato tutto quanto, "cielo e terra" e quindi anche l'uomo con la terra suo "habitat naturale". Con ciò i cristiani riconoscono il diritto di Dio sul mondo e credono nella promessa che il Creatore è anche per sempre conservatore e salvatore. Chi proclama questo enunciato della fede fa distinzione tra Creatore e creatura e, nella sottomissione a Dio, mantiene unite le due realtà. La creazione è soggetta al divenire e ad esser distrutta... Perciò noi preghiamo per la conservazione del mondo e speriamo nella redenzione di tutte le creature. I cristiani pregano con il salmista: Del Signore è la terra e quanto contiene, l'universo e i suoi abitanti. È lui che l'ha fondata sui mari e sui fiumi l'ha stabilita" (Sal 24:1–2).'

122 The Hebrew expression is עָבַד ('abad). See p. 204 below.

123 The Hebrew expression is שָׁמַר (šhamar). See p. 204 below.

124 See Presbyterian Church USA, 202nd General Assembly, Report *Restoring Creation for Ecology and Justice* (1990).

125 Pontifical Council for Promoting Christian Unity, *Ecumenical Directory*, 215.

126 Cf. V. Guroian, 'Toward ecology as an ecclesial event: Orthodox theology and ecological ethics' in *Communio* 18/1 (Spring 1991), pp. 89–110.

127 Cf. I. Zizioulas, *Il creato come eucaristia* [*Creation as Eucharist*] (Magnano: Edizioni Qiqajon, 1994). Cf. also pp. 231–233 below.

128 Inter–Orthodox Conference on Environmental Protection, *Conclusions and recommendations. Orthodox Churches and the Environment* (Crete, November 1991).

129 Cf. *Joint Statement of Orthodox Primates* (Constantinople, 15 March 1992).

130 World Council of Churches, *Final document of the World Convocation on Justice, Peace and the Integrity of Creation* (Seoul, 5–12 March 1990).

4

Christian Vision of Creation

But keep, O Lord, our journey through this life free also from storm and hurt unto the end. Send down refreshing rain upon the places that have need of it; gladden and renew through it the face of the earth, that it may delight in the refreshing drops and become green ... Bless, O Lord, the fruits of the earth, keep them for us free from disease and hurt, and prepare them for our sowing and our harvest ... Bless now also, O Lord, the crown of the year through Thy goodness for the sake of the poor among Thy people, for the sake of the widow and the orphan, for the sake of the wanderer and the newcomer and for the sake of all who trust in Thee and call upon Thy Holy Name.

Egyptian Liturgy of St Mark

4.1 Saint Benedict and Saint Francis

Long before the current ecological movement developed, saints taught respect for all of God's Creation. In fact, we can find two examples of the Christian attitude toward nature in Saint Benedict and Saint Francis. The true and good ecological spirit has been present for centuries in the Christian tradition. Saint Benedict represents the more active and practical aspect. The monks at Monte Cassino followed the rule of work and prayer (*ora et labora*), and learned to cultivate the land for intensive production without degrading the environment. In a community context, the good ideas of conservation and preservation were put into effect. Saint Benedict's approach was an example of an application of the Gospel passage: 'Seek first the kingdom of God and his righteousness, and all these things will be given you besides' (Mt 6:33).

Saint Francis, on the other hand, represents the aspect of praise and contemplation, exemplified by his Canticle of the Creatures: 'All praise be yours, my Lord, through our Sister Mother Earth, who sustains us and governs us, and produces various fruits with coloured flowers and herbs.' Saint Francis felt a natural — not pantheistic or intellectualist — brotherhood with every creature and every environmental event (for example: wolves, fire, water and even death). This type of perception does not seem to contrast the work of the Benedictines at all. Saint Francis, in fact, recommends not cutting down entire trees, but rather some of the branches in order to allow the tree to live and man to use its wood. As stated earlier, in 1979 Saint Francis was declared the patron saint of the environment by Pope John Paul II.[1]

Modern experimental science was made possible by the Christian philosophical atmosphere of the High Middle Ages. Modern science is the genuine product of a Judeo–Christian view of the world, which has its origin of inspiration in the Bible and in the doctrine of the *Logos*.[2]

This comes out of the fact that the Judeo–Christian vision of creation is diametrically opposed to the series of eternal returns that are found in ancient non–Christian and pre–Christian belief systems. The characteristic of pagan cosmogonies is a presentation of the inescapable birth–death–rebirth cycle, with no beginning or end and an essential lack of any meaning. In such a cyclic and eternal view of the universe, science could not make progress.[3] This is because science needs to be able to investigate the beginning of the processes of the universe. An adequate notion of time is crucial for developing differential calculus and integrals.

Precisely the progressive, linear world view deriving from Christian doctrine led to the growth of science, as well as other aspects of the human journey. The Christian faith, then, despite being primarily connected with the idea of eternal life, has a true effect on the world in which it is found. Christianity led to both material and spiritual effects, since 'The Gospel has truly been a leaven of liberty and progress in human history, even in the temporal sphere, and always proves itself a leaven of brotherhood, of unity and of peace.'[4]

Even though a talent for science was certainly present in the ancient world (for example in the design and construction of the Egyptian pyramids, and in the discovery of gunpowder and magnets in ancient China), the philosophical and psychological climate was hostile to a self–sufficient scientific process. Science therefore experienced incomplete 'births' in the cultures of ancient China, India, Egypt and Babylon. It also failed in its full realization among the Maya, Inca and Aztec peoples in the Americas. Even though ancient Greece came much closer to reaching a continuous scientific undertaking than any other ancient culture, science did not wholly emerge there either.

Despite their enormous advances in physics, medicine and mathematics, science was not fully birthed by the Muslim heirs of Aristotle. These 'stillbirths' of science can be connected to a primitive understanding of the cosmos as having eternal cycles in a necessary universe. The psychological climate of these ancient cultures often implicated a lack of hope or merely settling for what they knew, and in both cases there was a failure to believe in the existence of God the Creator, and an inability to produce a self–sufficient scientific undertaking.[5]

It is important, at this point, to emphasize that science grew within and was born out of the Middle Ages, in relation to a Christian view of the cosmos. Science and technology developed from philosophical roots that emerged from the Christian understanding of a rational and contingent cosmos, created out of nothing (*ex nihilo*) and with time (*cum tempore*) by God the Creator. The 'womb' of science was the Middle Ages, during which theology, science and philosophy worked together harmoniously. After the medieval period, however, this relationship between science and theology was fractured by Descartes, Kant and the Enlightenment.[6] Sciences little by little lost their reference to God the Creator. The moral criteria for judging the technological applications of science are no longer connatural to science and to technology. For this reason, technology has developed without the moral instruments to identify the grave consequences of abusing nature. This situation was verified in practice during the industrial revolution:

> The God and ethics of 'practical reason' were bound to become a matter of self–centered practicality in full accord with the self–centeredness imposed by Kant on thinking. This subjective ethical practicality

found its supreme sanction in the image which evolutionism paints of man. In that image ethics is reduced to man's practical responses in his struggle for survival.[7]

4.2 The ecological challenge to theology

There are several authors, often in the secular realm, who blame Christianity for the ecological problems in a society dominated by technology. They affirm that the roots of our difficulties are profoundly religious, citing the idea that the will of God is for man to make use of nature to his own advantage, according to the words of the Book of Genesis: 'Fill the earth and subdue it' (Gn 1:28). Lynn White was the first to sustain this thesis.[8]

According to White, Christianity inherited the refutation of cyclical time from Judaism (this is true), and substituted cyclic time with a linear concept of history, leading to a consequent rise in the hope of ever–growing and limitless progress; a rigorously anthropocentric world view must be added to all of this, derived from considering *homo sapiens* to be the *imago Dei*, which, according to White, makes people free to use and abuse a world whose dominion they can flaunt in the name of God. Modern science emerged from this context as an extrapolation of the Christian theological understanding of nature, effectuating the unconditional surrender of the earth to the will of man. Another negative element of Christianity, according to White and other more extreme authors, is the radical dualism between a being (man) who no longer considers himself an integral part of nature, and nature itself, whose mysterious vengeance does not delay in responding. This seems to be an echo of the Marxist dialectic.

For White, science and technology are so strongly saturated by a so–called 'Christian arrogance' that they cannot be relied upon to heal the crisis; the Christian faith itself, conveniently recycled (for example, according to the model of Eastern Christianity, much more supernatural and respectful of nature than the Western branch), must initiate a renewed mentality conversion. While White demands a mentality change from within Christianity, other more extreme secularist exponents reject the Christian perspective altogether. It must be stated that the ideas rejected by White and others are not Christian at all. In particular, the notion of continual and limitless progress is characteristic of Hegel or Marx, and of neo–Darwinism, which, among other things, do not account for the reality of original sin.[9]

Lynn White's thesis was re–proposed much more unilaterally by Carl Améry,[10] and a partially modified version of it was sustained by the Australian, J. Passmore, who identified the true root of present evils in the union of Greek culture with Christianity.[11] Later, things shifted toward Deep Ecology, so named because it asks deeper questions concerning 'why' and 'how' and thus is concerned with the fundamental philosophical questions about the impact of human life as one part of the ecosphere, rather than with a view of ecology as a branch of biological science. The core principle of deep ecology as originally developed is Arne Næss's doctrine of *biospheric egalitarianism* — the claim that, like humanity, the living environment as a whole has the same right to live and flourish.[12]

These views are very far from Christianity and, in fact, from any human truth, and readily lead to pantheism, in which individuality is lost in the haze of totality. They readily harmonize with some Eastern religions, especially Hinduism, Buddhism and Taoism, and encounter the favour of those who, like F. Capra, embrace New Age spiritual attitudes.[13]

Such types of eco–ideology are often mixed together with leftist politics and extreme feminism. In this approach, they misguidedly seek the possibility of definitively overcoming the framework of the ecological question provided by the anthropocentric tradition, typically inspired by a patriarchal — and therefore authoritarian and violent — logic, incapable of guiding humanity toward a harmonious and balanced relationship with natural realities.[14]

The eco–ideologies are very critical of traditional Western cultures, and in particular of historical Christianity. They expand and stretch the notions of sexism in the domination of men over women, and of racism in the domination of masters over slaves, into speciesism: the domination of the human species (in practice, men) over animals and the earth.[15]

Walter Kasper notes that:

> In effect, biblical faith in creation constitutes part of the spiritual presuppositions of the modern natural sciences and of the technological development that they have made possible, since it was the biblical distinction between Creator and creature which demythologized the world and saw it no longer as divine, instead identifying it as God's creation. Additionally, a rational world is created by a rational God. The environmentalist movement has shifted away from the Judeo–Christian vision. In separating from biblical anthropocentrism, the environmentalist movement sustains not just a nearly mystical unity between man and the world,

but also a sort of pantheistic definition of the relationship between God and the world. This vision of some environmentalists offers a new challenge for theology.[16]

It must also be confirmed, in response to environmentalists, that the awareness of being creatures should lead people to a religious reverence toward God for the whole of creation. The destiny of creation is entrusted to man, insofar as he is the image of God (Gn 1:28); he is absolutely not an arrogant and despotic lord: he is just a superintendent and manager, administrator and steward.

The position he has been given does not authorize him to plunder, radically alter or destroy the reality with which he has been entrusted and to which he is connected (Gn 2:7). In fact, it obliges him to promote it, defend it, and lead it to fulfilment; the stewardship entrusted him implicates wisdom, prudence and faithfulness, and excludes egoism, greed and lack of reflection. The passage of Genesis 2:15 (perhaps three centuries prior to Genesis 1:28) already spoke about the 'care for' the earth, and not just about man's utilization of it.

The same context of Genesis 1:28 indicates very clearly that only God is the Lord; as such, He founded creation on the basis of several natural laws that escape human jurisdiction and which man, like the other creatures, must respect. It is the same faith in creation, making the absolute sovereignty of the Creator known to man, which places limits on human dominion over the earth. When man abuses the position he has received, the consequences are just as much his own debasement as that of his living environment.

In most cases, environmentalists do not take into consideration the Christian dogma of original sin, which has secondary effects on man's intellect and will. We must also keep in mind the Christian teaching on personal sin. There is a notable proposal which suggests that it was the ancient Romans, with their business–oriented mentality, who translated the anthropocentric idea of Greek thought into practice and laid the basis for the attitude of dominion and possession which is still so rooted in Western culture today.[17]

Still others, such as Max Weber, connect Protestantism with liberal capitalism.[18] It is interesting to note the differences between the understandings of the person–nature relationship in the Catholic tradition and in the Protestantism of the Reformation.[19] In Luther's opinion, as in much of Protestant theology, the kingdom of God and the kingdom of the world are seen as being in tension, in a certain form of reciprocal contrast or antithesis. This position is connected to Luther's idea (which is later also taken up by Jansen) that human nature is totally corrupted by original sin. There is therefore an opposition between human nature and grace, which has several consequences for the relationship between humankind and the natural environment. On the other hand, when human nature and grace are seen together in synthesis as in the Catholic tradition (particularly evident in the Greek Fathers of the Church and the Thomist perspective), the relationship between humanity and nature is understood in a more positive way: the general disposition is of cooperation with nature rather than opposition to it.

The Catholic tradition puts greater emphasis on the principle of the Incarnation and the principle of sacramentality than the Protestant tradition. The Protestant tradition (in environmental theology) puts the focus on the idea of service or of a strong sense of responsibility for nature and toward other present and future members of humanity. It also has a tendency toward individualism.

This approach is insufficient. A more Christological position is necessary, which comprehends the connection between creation, the Incarnation, the Paschal Mystery and eschatology. The ecological criterion must be based on love for Jesus Christ in nature, and love in Christ for other present and future persons. In its moral impulse, the Catholic position highlights a morality based on 'being,' or rather, with an ontological basis. On the other hand, the Protestant view proposes self–sufficient morality, without metaphysical roots.

4.3 Principles of environmental theology

The Christian theology of creation directly contributes to the solution of the ecological crisis, affirming the fundamental truth that visible creation is itself a divine gift, the 'original gift,' which creates a 'space' for personal communion. Effectively, a correct Christian ecological theology is found in the application of theology to creation. The term 'ecology' combines the two Greek words, *'oikos'* (house) and *'logos'* (word): the physical environment of human existence could be seen as a sort of 'habitation' for human life. Considering that the interior life of the Holy Trinity is a life of communion, the divine act of creation is the totally free production of partners who can share in that communion. In this way, it can be said that the divine communion has now found its 'habitation' in the created cosmos. For this reason, it is possible to speak of the cosmos as a place of personal communion.[20] At the same time, it is clear that theology will not be able to provide a technical solution to the environmental crisis; nonetheless, theology can help us see our natural environment as God sees it, as the place of personal communion in which human beings, created in the image of God, must seek reciprocal communion and the final perfection of the visible universe.[21]

The Christian view of creation is the core of this treatise and is of fundamental importance for the Christian foundation of a new responsibility toward the environment. Philosophical realism also has its place in considering the theology of the environment. Realism is an instrument of dialogue between science and faith.[22] Realism and theological language are necessary for us to develop a correct understanding of the environment. It is important to consider the cosmos from the scientific, philosophical and theological points of view, seeing its relationship with anthropology and thus avoiding the error of cosmocentrism. The realist perspective is also necessary to establish the basis for moral action with respect to the environment.

4.3.1 Creation and Revelation

In Wisdom literature and in the Psalms, the origin of creation is recalled in the creative Word of God (cf. for example Ps 33:9). The world comes to be as solidly founded on God. People find assurance and support in this. Wisdom exalts the beauty and order of creation as a testament to the greatness of God. The created world must turn into joy and glorification of the Creator (cf. Ws 13:1–5; Sir 42:15–43; Jb 12:7–9), because in it the goodness and wisdom of God become apparent (cf. Ps 8; 104). It is the breath of God, the spirit of wisdom and goodness which fills the earth (Ws 1:7; 8:1): 'For you love all things that are and loathe nothing that you have made; for what you hated, you would not have fashioned. And how could a thing remain, unless you willed it; or be preserved, had it not been called forth by you? But you spare all things, because they are yours, O Lord and lover of souls' (Ws 11:24–26). The love of God for His creatures helps us understand creation as a relational reality.[23]

Over the past century, faith in creation has been primarily reduced to the affirmation that everything which exists is due to divine causality. There has been a tendency to see the content of the Christian faith as a response to the word of revelation pronounced throughout salvation history.

There has always been, however, a conviction that not only salvation history, but also creation itself constitutes a context of word and event in which God expresses Himself and turns toward man. Since early times, beginning with Tertullian and Saint Augustine, the great theologians have spoken of a 'double' book of Divine Revelation: creation and Holy Scriptures. The Scholastics developed the doctrine of the world of things as *imagines et vestigia Dei*, and used concepts such as sacrament of nature or of creation.[24] The metaphysical traces or imprints, such as unity, beauty, truth and contingency, are found inscribed in creation just as labels are found on the 'creations' of clothes designers. But the full meaning of creation is only intuited through supernatural revelation. As a theoretical foundation for a Christian response to the ecological challenge, therefore, we must turn to a renewed creation theology.

There is a limited meaning in which creation carries some form of God's Self–revelation. Saint Anselm of Canterbury wrote: '*Uno eodemque Verbo dicit seipsum et quaecumque fecit*' (In the one and the same Word, God proclaims Himself and what He has done).[25] Seen in this way, created things are '*verba in Verbo et de Verbo*' (words in the Divine Word originated by the Divine Word). Creation with all that it contains is, in some analogical fashion, an expression, a symbol and a sacrament of God's action. Consequently, there are not merely obscure allusions to God in creation, but in it He reveals Himself, as Saint Paul affirmed in his letter to the Romans (Rm 1:19). Or, in the words of Saint Bonaventure: 'Everything that God does, He does in order to manifest Himself.'[26] In a certain limited sense, creation also prefigures God's gift of Himself. God is He Who, according to Sacred Scripture, gives life to all, provides food and drink, and brings rain and sunshine upon the just and the unjust to show His love, His concern and His gift to humanity.

It must be emphasized that there are various degrees of intensity in this gift. In the human realm, a handshake is less intense than the expression of spousal love, which in turn is a less intense level than sacrificing one's life for Christ and His Gospel. The intensity of God's manifestation is found in a certain measure in creation, and in a much greater measure in salvation history, which culminates in the sacrificial offering of the Son of God upon the altar of the Cross.

Creation itself, furthermore, is already a first modulation of the Word of God. In a certain analogous and limited sense, creation is a 'sacrament,' a sign and efficacious means, of God's Self–revelation and His gift of Himself. In creation, God manifests Himself in a mediated way. It is necessary to make these clear distinctions in order to avoid ontologism and other steps toward pantheism.[27] A distinction must be made between natural and supernatural revelation. Theodoret, Bishop of Cyrrhus in Syria, highlighted the enormous difference between God's revelation in nature and in His Son made flesh:

The Incarnation of our Saviour represents the greatest fulfilment of divine solicitude toward man. In fact, neither heaven nor earth nor sea nor sky nor sun nor moon nor stars nor the entire visible and invisible universe, created by His word alone or rather brought to light by His word in accordance with His will, indicate His incommensurable goodness so much as the fact that the Only Begotten Son, He Who subsists in God's nature, reflection of His glory, imprint of His substance, Who was in the beginning, was with God and was God, through Whom all things were made, after having taken on Himself the nature of a servant, appeared in human form. By His human form He was considered man, was seen on earth, interacted with men, bore the burden of our weaknesses and took upon Himself our ills.[28]

The declaration of the German Bishops' Conference also considers the theology of the cosmos as gift:

If we consider the world a creation of God, it appears much differently to us and becomes new. It is the gift of a God who loves ... Accepting the reception of the world as a gift: this also gives us a new love for the world and for living things. They become precious to us, a gift to give to others, a sign and symbol of the goodness of God.[29]

4.3.2 *Createdness of the world*

In the passage of Genesis 1:1, we find the first words of the Bible: 'In the beginning, when God created the heavens and the earth...' In this phrase, 'create' (*bārā*) is a precise theological verb whose subject is almost exclusively God. The word itself cannot be an absolute proof of *ex nihilo* creation, because there are some exceptions (Jos 17:15–18,

Ez 23:47), but all of Genesis 1 indicates *ex nihilo* creation. Even reason, on the basis of a logical–semantic analysis of the biblical text (supported by the most recent conclusions of philosophers of science), can affirm this truth. This conflicts with the beliefs of nearby contemporary peoples of the Middle East who followed other, erroneous world views.

The verb 'create' is clarified through the verb 'speak' (*dabar*). The first book of Genesis testifies no less than seven times — once before every individual act of creation — that God spoke, and that in function of this speaking the world became a reality. This notion of 'speaking' implies revelation, and highlights the free and personal nature of God in His act of creation. God takes the initiative. The verb 'bless' (*baruk*) adds something more than goodness to 'creating.' By blessing, God promises fertility and abundance to the creatures. Blessing remains and persists despite all human errors.

God is absolutely free in His creating. Therefore, He does not create by necessity. With creative spontaneity, He founds a reality distinct from Himself and He freely provides it with its specificity. The world is contingent in its existence because God was free to create a world or not. The form in which God created the world is not a necessary form, but rather a contingent one insofar as it depends on a single choice (among many possible choices) that God made.

Creation, therefore, is distinct from God. If the world is the product of the creative word of God, then it is clearly distinct from God in its very nature. God's freedom also closes the door to pantheism. The cosmos is good and there is no danger of collapse, as was feared in the ancient pagan understandings based on inherent instability. The cosmos is unique and it is a whole. God is supremely rational. Therefore, creation reflects a rationale and so it cannot be used arbitrarily. There is logic in the natural and revealed laws concerning the use of creation.

The discussion on createdness has two important aspects. Firstly, it is an action that occurred once and for all, which is to say *creatio ex nihilo* and *cum tempore*. Secondly, it is a beginning that lasts and develops throughout history, in the design of God's Providence. If we ignore the initial moment, the danger of pantheism becomes reality; if we instead ignore Providence, the danger of deism becomes reality. The cosmos was not only created, but it is also a creature.

Many secular environmentalists who believe in only a vague type of Supreme Being, but not in Christianity, fall into pantheism (where nature becomes an object of worship and adoration) or deism. Instead, Providence is the basis for understanding human beings as stewards, who participate in the Paternity of God and in Providence.

God transcends His creation, but follows it closely in an immanent way. At this point, the question of the 'positive way' (connected to immanence), the 'negative way' (connected to transcendence) and the 'eminent way' comes into play: this last way affirms that the very perfections of caused realities exist superabundantly in God.

4.3.3 Place of the human person

We reject Kant's absolute anthropocentrism in which man is imprisoned in both a subjectivism and an agnosticism with respect to God, the cosmos and the human soul. Indeed extreme anthropocentrism leads to cosmocentrism, simply because once man's reference point in God is removed, his ideas become like flotsam and jetsam floating towards any kind of weak ideology. Man is part of the cosmos; but eliminating the distinction between man and the cosmos leads to cosmocentrism. This is also the danger in R. Dawkins' neo–Darwinism, in which man is only the carrier of genetic wealth.

Let us return instead to the biblical understanding of man's place and follow Christian Christocentrism. In the first chapter of Genesis, man and woman appear to be the culmination point towards which the history of creation tends, step by step. In the second chapter of Genesis, human beings appear as the centre around which God creates His world. In both cases, man and woman are always seen in reference to God. Man is the superintendent or steward of creation.

a) The Creator's decision

In Genesis 1:26, the creation of man begins with an explicit, divine decision. The expression 'Let us make man...' expresses the majesty of God. All other creatures were created through the 'word.' In Genesis 1:3, 6, 9, 14, 20, 24, the expression 'God said' is found six times. This means that God set Himself as a dynamic force and as the meaning and final end of the world. However, something much greater occurs when the Sacred Text uses the word *bārā* three times for 'create' to describe the action of the Creator in forming man; God is described almost as starting with a new decision from our perspective (*quoad nos non quoad Deum*). This is intended to clearly highlight, with as much emphasis as possible, the fact that man's creation depends in an entirely special way directly on God Himself, and that this work brings the creative act to its apex and true end. The creation of the human person is something special and unique, to be distinguished from the creation of the animals. The fact that the expression 'created' is used three times in Genesis 1:27 could have various meanings and lead to different interpretations. First of all, it indicates that the human being is the image of the Holy Trinity. Secondly, it shows that God is responsible for the creation of the human being's body and soul, and for the relation and distinction between men and women.

b) Man is in the image of God

In the Book of Genesis 1:26–27 we read: 'Then God said: "Let us make man in our image, after our likeness. Let them have dominion over the fish of the sea, the birds of the air, and the cattle, and over all the wild animals and all the creatures that crawl on the ground." God created man in his image; in the divine image he created him; male and female he created them.' Here we find the basis considering the human being as the image of God. Man is considered a 'steward' in the Western Christian perspective, whereas in the Christian East he is conceived of as a 'priest.' In summary, it can be said that the first chapter of Genesis is at the basis of the Western approach, whereas the second chapter is related to the Eastern approach.

As a creature, man is a *living* being (Gn 2:7), a quality apparently shared by all animals. But human createdness is different in a specific and unique dimension that animals do not have: in addition to having the 'breath of life,' which is the light of self–consciousness, men and women are 'images of God.'

Being the 'image of God' means that men and women not only 'exist,' but are capable of a relationship with God, if God wishes. On the one hand, then, man is connected to his world ('out of the clay of the ground'), and on the other hand he is open ('image of God') to relating with God. Being the image of God is the basis for a relationship of intimacy with God. It is important to point out that being the image of God does not only refer to the human soul, but also to the human body.[30]

Human beings in their entirety were created in the image of God. This perspective excludes interpretations which place the *imago Dei* in one aspect or the other of human nature (for example, in righteousness or in the intellect), or in one of their qualities or functions (for example, sexuality or dominion over the earth). Avoiding both monism and dualism, the Bible presents an understanding of the human being in which his spiritual dimension is seen together with his physical, social and historical dimensions.[31]

Far from encouraging an unbridled and anthropocentric exploitation of the natural environment, the theology behind the *imago Dei* affirms man's crucial role in the realization of God's eternal abiding in the perfect universe. Human beings, by God's design, are the administrators of this transformation for which all of creation yearns.[32]

c) Goodness in the world

In the Book of Genesis 1:4, 10, 12, 18, 21, 25, it is affirmed six times that 'God saw how good it was.' This goodness is in the ontological order. This goodness must not be reduced to a moral dimension or to a useable dimension understood in a pragmatic sense; it must include the capacity of creatures to reflect the glory and perfection of God. The affirmation of goodness does not concern only spiritual creatures, but also material ones in their various forms. The determination of goodness concerns the act of creation as such. This goodness is articulated in a hierarchical form.

Creation is only 'good' before the creation of man. Not everything that God made is good in the same way. That which is merely 'good' becomes 'very good' to the degree in which it attains its greatest fulfilment, made possible through man. In Genesis 1:31, after God created man and entrusted him the responsibility of the earth, it is declared: 'God looked at everything he had made, and he found it very good.' Creation is 'very good' only after God places a central reference point in it: man, through whom it all becomes a meaningful whole with a unitary and comprehensive order. This idea saves us from cosmocentrism.

From this argument, it can be deduced that the value of nature does not consist solely in its utility for man. Nature constitutes a value in itself, in reference to the Creator, as a hymn of praise to the Creator. However, the subhuman world attains its fullest meaning only in reference to man. At the same time, man attains his fullest meaning in his relationship with God. The idea of the equality, equivalence and autonomy of all creatures does not accord with the faith of the Church.[33] There is a hierarchy of participation and solidarity in the cosmos.

d) Fill the earth and subdue it

The reference point for this discussion is Genesis 1:28: 'God blessed them saying: "Be fertile and multiply; fill the earth and subdue it. Have dominion over the fish of the sea, the birds of the air, and all the living things that move on the earth."' The word 'subdue' as translated in this context corresponds to the Hebrew expression *kâbâs*, which means 'taking possession of a territory.'[34] With the divine blessing, humanity receives the ability to generate and multiply to the point of filling the earth. This indicates the idea of fecundity at the beginning of the process which leads toward pleroma, or fullness.

The other key phrase, 'have dominion,' corresponds to the Hebrew *râdâ*, which means 'to herd, conduct, guide, govern.'[35] It is thus more attenuated than the modern understanding of 'dominion.' The territory and the animals were entrusted to man. But that entrusting occurred through a blessing which was given to man insofar as he was in the image of God. This means that the relationship between man and the territory and the animals must be in conformity with the Providential will of God. It is therefore not an arbitrary relationship, and human beings cannot arbitrarily make use of what has been entrusted to them. As N. Lohfink proposed:

> this blessing does not at all legitimize the destruction of entire families of animals on the various continents, of marine organisms, or of countless avian and insect species, in the name of man's transformation of the face of the earth ... This blessing means the opposite.[36]

It is important to read the passage of Genesis 1:28 in relation to the passage of Genesis 2:15. Sacred Scripture must always be read in a simultaneously analytical–comparative and synthetic way.

e) Cultivate and care

Genesis 2:15 presents this key phrase: 'The Lord God then took man and settled him in the garden of Eden, to cultivate and care for it.' The garden of Eden should not be considered an intangible, magical place, a 'virgin' forest not to be entered. It is a world entrusted to man's care. It is a perfect and complete garden, the reality and symbol of

God's absolute gift and unconditional promise to man. The symbol is bound to reality. It is a part of the cosmos, and in this sense it is a symbol of totality on the basis of the *pars propter totum* principle.[37] God does not seek a collaborator to complete the garden, but rather a recipient to whom He can entrust it as His gift and His promise.[38] God's gift is complete and perfect: 'God looked at everything he had made, and he found it very good' (Gn 1:31). This perfection is also expressed by the Lord's rest: 'Since on the seventh day God was finished with the work he had been doing, he rested on the seventh day from all the work he had under-taken' (Gn 2:2). The work of cultivating and caring is the grateful experience of a gift, a rejoicing with God in His creation, as Psalm 104:31 sings: 'May the Lord be glad in these works!' The human response is appreciation for the gift of creation.

The biblical task of working in and on the created world should be understood in the sense of reproducing 'divine work.' The Sabbath rest places limits on man's interaction with the world, an interaction of work by which he can give shape to and modify it. It provides an open space in which man can always newly orient himself in accordance with the fundamental image of God. Therefore, any permission for arbitrary domination of nature — any reck-less exploitation or destruction — is excluded by principle. Man's tasks are instead the regulation of order and the reduction of conflicts, aiming to develop the positive aspects while keeping the destructive forces of nature in check.[39]

The two verbs used here for 'cultivate' (in Hebrew *'abad*) and 'care' (in Hebrew *šamar*) evoke a religious atti-tude, because *'abad* is not only 'agriculture,' but also indicates the service of worship through a relationship with God, while *šamar* expresses both God's loyalty to man and man's loyalty to God, thus evoking the covenant. The attitude of caring is not constituted simply by the exercise of power, but of recognition and praise. In fact, one cares for something that is valued as a precious and expensive good. Man's 'care' for the world is inseparable from man's service to God.[40]

f) Man's autonomy

According to Genesis 2:19 and the following passages, the Lord God made the animals from the ground and presented them to man to see what he would call them; and man gave the animals names. Conferring a name, in the Hebrew mentality, does not mean that man simply makes up some words and applies them to the individual animals. Giving a name is a sign of a right of sovereignty, a role of dominion, and for this reason Genesis 2:19f is very similar to Genesis 1:28. Man welcomes the animals just as God made them, but, giving them a name, they become part of his own world. A demythologizing occurs in this process: the animals lose any divine quality, and are ordered to man as part of his living space to be organized freely and responsibly.

God, in His wise Providence, wishes to bring the universe to its fulfilment as much as possible through creatures themselves. For this reason, He gave things a natural order in which creatures depend on one another and through which their being is preserved. 'Now it is a greater perfection for a thing to be good in itself and also the cause of goodness in others, than only to be good in itself. Therefore God so governs things that He makes some of them to be causes of others in government; as a master, who not only imparts knowledge to his pupils, but gives also the faculty of teaching others.'[41] In this discussion, the appropriate distinction must be made between autonomy 'from' and, more positively, autonomy 'for,' which leads to a perspective of participation and solidarity. Man, created with intelligence, is made capable in a particular way of participating in the divine governing of the world, and of advancing the divine plan through history. This is the meaning of man's autonomy, which is a participation in (not a separation from) the divine economy.

Autonomy must not be understood in an absolutistic or Enlightenment–inspired perspective (along the lines of Kant), which leads on the one hand to individualism and to collectivism on the other. It must not be seen as a human pretension to snatch an arbitrary relationship with the world, ignoring its laws and structures, and setting aside its framework of meaning and values. Man's autonomy is instead founded on the rationality of reality as a whole. This formula of the rationality of reality expresses the conviction that the nature of the cosmos, the pre–established structures of the world, make it possible for man to develop a meaningful and fruitful existence. This relative and rational autonomy is a shield against pantheism. The cosmos as an interacting whole excludes the possibility of absolute autonomy.

4.3.4 Animals

A passage from the Book of Psalms describes the God's care toward animals: 'Your justice is like the highest mountains; your judgments, like the mighty deep; all living creatures you sustain, Lord' (Ps 36:7). This does not necessarily mean that animals receive the supernatural salvation that man does. It must be remembered that whereas animals were formed from the ground (Gn 2:19), human beings additionally received the breath of God (Gn 2:7). The alteration of the arrangement of the degrees of life in the pact with Noah (Gn 9) is an expression of the fact that at this point the competitive relationship between man and animals jumps out in a much sharper way. Man is allowed to kill animals for sustenance. However, he is not also conceded limitless discretional power. On the whole, man acquires a relatively greater position of responsibility, which also includes the rejection of evil. Prehistory clarifies that our real world is characterized by ruptures and wounds which man can no longer heal. Nature is taken to be conflictual and unforeseeable and is now marked by signs of 'hostility' (cf. Gn 3:17–19). Later Old Testament messages about the created order, for example in Psalms 8, 19 and 104, essentially overcome this conflict. Various regulations regarding the protection of animals (Dt 25:4; Ex 23:4f; Lv 25:7 and others) demonstrate the value of and immediate responsibility for animals, yet the killing of animals for sacrificial or alimentary aims and their use in production remain unquestioned. The regulations for feasts and the sabbatical year (cf. Ex 20:8–11; Dt 5:12–15; Ex 23:10f; Lv 25), as for the *Shabbat* in the creation story of the priestly tradition, preserve a promising spark which recalls the peace of creation and are the expression of a theology of joy for creation, which renounces maximal exploitation of it. This dimension becomes even clearer in the eschato-logical promise of the reign of messianic peace (cf. Is 11:6ff), which explicitly includes peace in creation (and therefore between man and animals).[42]

In Pope John Paul II's speeches, numerous references to the role of animals in creation can be found.[43] Clearly, as Christians we reject any type of cruelty to animals. This cruelty becomes more serious according to the place that the animal has in the evolutionary hierarchy, which means: the more evolved the animal is, the more it can 'suffer.' Obviously it is licit to kill animals for protection (for example, from a snake or a scorpion), as well as for food and clothing provided that cruelty is avoided. Forms of sport which involve cruelty to animals (such as hunting) are inappropriate for a Christian. Those who are cruel to animals might be uncivil with other people as well.

Scientific experimentation on animals must be limited as much as possible, also because human physiology is different from animal physiology.[44] It must be affirmed, however, that the attitude of modern Western society is inappropriate in this regard: some people assert that animals think and love like we do — that they are equal to us. This is a reduction of human beings. For many, cruelty to animals is horrible, but abortion is acceptable! In the West there are cemeteries for dogs, but in Africa people who die of hunger are often not properly buried. The *Universal Declaration of Animal Rights* can be found in Appendix 6 of this book, and is an obvious example of reducing human beings to the dignity of animals by raising animals to the level of human dignity!

Some (including Catholics) ask if there is a place for their animals in heaven. Some, such as M. Damien, ask if animals can perceive transcendence and even talk about animal prayer.[45] Damien also very dangerously affirms that 'Christ also died for dogs.'[46] Saint Augustine instead teaches that the properties of the future world will be adapted to the immortal existence of the transfigured human body:

> the qualities of the corruptible elements which
> suited our corruptible bodies shall utterly perish,

and our substance shall receive such qualities as
shall, by a wonderful transmutation, harmonize
with our immortal bodies...[47]

One might ask, then, if there will also be plants and
animals in the new creation. Saint Thomas seems to offer a
negative response to this question; in his opinion, they are
incapable of receiving a renewal into incorruptibility, since
they

are corruptible both in their whole and in their
parts, both on the part of their matter which loses its
form, and on the part of their form which does not
remain actually; and thus they are in no way
subjects of incorruption. Hence they will not remain
in this renewal.[48]

Today, however, there is a tendency to ask if in the context
of the renewal of all creation there might not also be a place
for plants and animals in a new glorified kingdom of
matter. There seems to be no intrinsic reason to exclude
inanimate beings from the new creation. In fact, the unified
presence of angels, men and women, plants and animals,
and inanimate beings seems to some people more conso-
nant with the completeness of the new heavens and the
new earth.[49]

In the new creation, angels will be present because they
are immortal by nature and will enjoy the gift of God's
glory. Human beings will be resurrected by the power of
God, and animals and plants could be incorporated by
God into the renewal of material creation. This theory
would allow all of the old creation to be in some way
represented in the new. It is difficult, however, to speculate
on the continuity of a particular animal or a given plant
between this life and the future one, because these beings

do not have a spiritual soul. The ontological difference that exists between human beings and animals must always be remembered, since only man is created in the image of God.[50] The central and essential joy of Paradise will be the beatific vision of God face to face: there will also be a secondary joy in the company of Mary and of all the angels and saints of Paradise, including parents and friends, along with the renewed material universe.[51]

4.3.5 *The cosmos in light of the mystery of Christ*

Jesus Christ is the key to revealing the secret of creation. In the various Christologies of the New Testament, Christ is not only related to the second, new creation, but also to the first. According to St. Paul, Jesus Christ is the only Lord 'from whom all things are and for whom we exist' (1 Co 8:6). In the *Logos*, St John sees the universal and only mediator of creation: 'In the beginning was the Word ... All things came to be through Him, and without Him nothing came to be' (Jn 1:1,3).

The Christian view of creation rises above the most ancient religious systems in which the cosmos was considered eternal and cyclic with respect to time. In China, despite their differences, the Taoist, Confucian and Buddhist approaches have in common the idea of an eternal cosmos and a certain cyclic repetition throughout time. Similarly, the Hindu religions in India held that the cosmos was eternal and regulated by inexorable cycles.

In the pre–Columbus Americas, there were also elements of this sort. The gods of the Aztecs were personifications of various periodically changeable forces, and of natural phenomena. The cosmos was cyclic; the concepts of space, time and causality were missing. The Incas were prisoners of a cyclic idea of time. The Mayans also had a cyclic notion of time in which there was no beginning.

For the ancient Egyptians, the universe was like an enormous animal that gave origin to an organismic, rhythmic and animistic cosmogony. The Babylonians, Sumerians and Assyrians were closed within a cyclic and animistic understanding of the world, distinctly different from the Old Testament understanding. In the Aristotelian, Stoic and Epicurean cosmologies of ancient Greece, the universe was cyclic; matter and its processes were eternal.

In the neo–pagan philosophies of the Renaissance, German idealism and the New Age Movement, as in the attempts of modern scientists to exclude God from His creation, there is a return to a cyclic and eternal cosmos.[52] The Judeo–Christian outlook on creation is diametrically opposed to this whole series of eternal returns which are found in ancient pagan systems as well as in modern ones.

The doctrine that the eternal Word became incarnate of the Virgin Mary at a specific moment in history guarantees the uniqueness of Christ's coming and of His redemptive act. Another contribution of orthodox and dogmatic Christianity is a strong appreciation for time as actually experienced. The fact that the Incarnation happened at a precise moment in time, emphasized by the continual reference to Pontius Pilate in all professions of faith, increases the perception of the uniqueness of every moment and therefore of history. Since this uniqueness is inconceivable within cyclic recurrences, the Incarnation adds further emphasis to the linear perception of time, and this forms an integral part of the history of salvation begun in the Old Testament and fulfilled in Christ.[53]

> The eternal Father, by a free and hidden plan of His
> own wisdom and goodness, created the whole
> world. His plan was to raise men to a participation
> of the divine life. Fallen in Adam, God the Father
> did not leave men to themselves, but ceaselessly
> offered helps to salvation, in view of Christ, the
> Redeemer 'who is the Image of the invisible God,
> the firstborn of every creature.'[54]

Scotist Christology defended the thesis according to which
the Incarnation, as God's ultimate most original act, represents the supreme moment of His Self–manifestation and
that in this act, in some way, the will of creation is already
included. The Incarnation of the *Logos* thereby becomes the
true end of all the movement of creation and all of the rest
is merely preparation. The Christocentrism of creation is
illustrated well by the Christological hymn in Colossians
1:15–20, and in particular by Colossians 1:15–17:

> He [Christ] is the image of the invisible God, the
> firstborn of all creation. For in Him were created all
> things in heaven and on earth, the visible and the
> invisible, whether thrones or dominions or princi-
> palities or powers; all things were created through
> Him and for Him. He is before all things and in Him
> all things hold together.

The Christocentrism of the Christological hymn to the
Colossians is made up of three aspects. First, instrumental
efficient causality (*causa efficiens*) is indicated by the words
'in Him were created all things … all things were created
through Him …' (Col 1:16). The preposition 'through Him'
(δι' αὐτοῦ in Greek) means that Christ actively participates
in the act of creation. All things receive their existence and
redemption through His work. Second, one can examine
the phrase 'in Him all things hold together' (Col 1:17),

where the expression 'hold together' corresponds to the Greek συνέστηκεν. The preposition 'in him' (ἐν αὐτῷ in Greek), understood as a Semitism, also indicates instrumental efficient causality; this seems improbable in such a carefully developed piece of writing. Given the undeniable 'Alexandrian' feel of the hymn, it is possible to suppose that an exemplary causality is expressed here (*causa exemplaris*): Jesus becomes the Mediator between the invisible God and the visible world, because as the 'image' of the One, He is the exemplar for the other.[55] In effect, there are two elements to consider: first of all, the fact that Christ is the foundation upon which all of creation rests, as He sustains it and preserves its being through Providence; second, 'in Christ' is at the same time a soteriological affirmation: it means being brought, through grace, into the covenant with the Lord. The third aspect of Christocentrism in the hymn is finality (*causa finalis*), which is indicated by the statement that things are created 'for Him' (Col 1:16). In Greek, the preposition is εἰς αὐτὸν, which indicates a returning movement of things toward Christ. He is designated as the end (aim) of creation. This end is found in the question of redemption (v.19f): 'for in Him all the fullness was pleased to dwell, and through Him to reconcile all things for Him, making peace by the blood of His cross.'

The fact that creation and redemption are connected is seen when Christ appears as 'the firstborn of all creation' (v.15) and 'the firstborn from the dead' (v.18). Christ is the Mediator of creation and salvation. For St. Paul, creation and redemption are two aspects of a single great mystery: the recapitulation of all things in Christ, as can be read in the Letter to the Ephesians: God 'has made known to us the mystery of His will in accord with His favour that He set forth in Him as a plan for the fullness of times, to sum up all things in Christ, in heaven and on earth' (Eph 1:9–10).

The term ἀνακεφαλαιώσασθαι (*anakephalaiosasthai*) means 'to recapitulate' and is derived from κεφάλαιον, which has the meaning of 'sum total.' The process of gathering together again and reuniting all things in Christ implicates the lordship and majesty, both transcendent and immanent, of Christ over the entire cosmos. That kingship already exists; it must be brought to its fulfilment. The concept of kingship is important in the ecological discussion. God's saving plan, 'the mystery of his will' (Eph 1:9) concerning every creature, is expressed in the Letter to the Ephesians in a unique way: to 'sum up' or 'recapitulate' all things, heavenly and earthly, in Christ (cf. Eph 1:10). This image could also evoke the rod around which the role of parchment or papyrus of a *volumen* would be wrapped: Christ endows a unifying meaning to all of the syllables, words, and works of creation and history.[56] The first person to grasp and admirably develop this idea of 'recapitulation' was Saint Irenaeus, Bishop of Lyons and a great Father of the Church from the second century. Against any fragmentation of the history of salvation, against any separation between the Old and New Covenants, against any dispersion of divine action and revelation, Irenaeus exalts the one Lord, Jesus Christ, Who in the Incarnation ties into Himself the entire history of salvation, humanity and all of creation: 'the King eternal is raised up, who sums up all things in Himself...'[57]

The concept of πλήρωμα (*plèroma*), which means 'fullness,' is also important in this context, as can be read in the Letter of St. Paul to the Colossians: 'For in Him all the fullness was pleased to dwell, and through Him to reconcile all things for Him' (Col 1:19–20). God's fullness is found in Christ. In St. Paul's writings, the meaning of *'plèroma'* seems to be that all things were created in Christ, reconciled in Him, and find their eschatological fulfilment in Him. In a certain way, the fullness of all that exists can be found in Christ; everything depends on Christ and finds in Him its meaning and its very existence. Christ is the Head of the cosmos and all things are already recapitulated in Him, but not yet fully. They are recapitulated in an incipient way, and the fulfilment of the reconciliation of all things in Christ takes place in the realm of history, through the Church.

Furthermore, it must not be forgotten that both the hymn in the Letter to the Colossians and the eulogy in the Letter to the Ephesians cannot be interpreted without taking sin into account. This coming together of all being in Christ, the Centre of time and space, occurs progressively through history by overcoming the obstacles and resistance of sin and of the Evil One.[58] In order to illustrate this tension, Irenaeus makes use of the contrast between Christ and Adam, as already discussed by St. Paul (cf. Rm 5:12–21): Christ is the new Adam, the Firstborn of faithful humanity, Who welcomes in love and obedience the redemptive plan that God has laid out as the essence and aim of history. Christ must therefore cancel out the devastating works, the horrible idolatries, the violence and every sin that the rebellious Adam has sown in the course of human history in the created world. With His full obedience to the Father, Christ opens the era of peace with God and among men, reconciling human enmity in Himself (cf. Eph 2:16). He 'recapitulates' Adam — in whom all of humanity is identified — in Himself, transfiguring him into a child of God, bringing him to full communion with the Father. Precisely through His fraternity with us in flesh and blood, in life and death, Christ becomes the 'Head' of saved humanity.[59] St. Irenaeus further writes: 'Christ has recapitulated in himself all of the blood shed by all the just and by all the prophets who have lived since the beginning.'[60]

4.3.6 *The mystery of evil and sin*

After the fall of man, there is a constant danger of man's abuse of power. Original sin, in fact, damaged the relationship between God and man, between man and the cosmos, and the relationships of human beings with one another. Oftentimes, the approaches of secular environmentalists lack any consideration of sin and evil.

After man's original Fall, the world — as it had been entrusted to him — in a sense came to share his lot. Sin not only broke the bond of love between man and God and destroyed the unity of mankind, but it also disturbed the harmony of all creation. The shadow of death came down not only on the human race but also on everything that by God's will was meant to exist for man. However if we speak of the cosmos sharing in the effects of human sin, we also know that it has been given a share in the divine promise of the Redemption. The time for the fulfilment of this promise for mankind and for all creation arrived when Mary, by the power of the Holy Spirit, became the Mother of the Son of God. Christ is the firstborn of creation (cf. Col 1:15) and He came in order to embrace creation anew, to begin the work of the world's redemption, to restore to creation its original holiness and dignity.[61]

Good and evil must be considered in light of the redemptive work of Christ. This redemptive action, as St Paul leads us to infer, involves all of creation in the diversity of its elements (cf. Rm 8:18–30). Nature itself, in fact, just as it is subjected to the degradation, devastation and fragmentation of its meaning caused by sin, thus participates in the joy of liberation carried out by Christ in the Holy Spirit. The full actualization of the Creator's original plan thus comes into sight: a creation in which God and man, men and women, humanity and nature are in harmony, dialogue, and communion. This plan, disrupted by sin, has been recovered in an even more marvellous way by Christ, Who is mysteriously but efficaciously enacting it in the current reality, waiting to bring it to completion. Jesus Himself declared that He is the fulcrum and focal point of this plan of salvation when He affirmed: 'And when I am lifted up from the earth, I will draw everyone to myself' (Jn 12:32). John the Evangelist presents this work precisely as a sort of recapitulation, a gathering into one the dispersed children of God (cf. Jn 11:52).[62]

The understanding of Christian Scripture and Tradition is contrary to the notion that evil is an integral or necessary part of the nature of the cosmos, or that good and evil are two faces of the same ultimate reality. This is the monistic understanding. The Christian understanding opposes dualism as well, for example Manichaeism, whereby good and evil are two equal and opposite forces. Christianity rejects the false notion that evil exists only because of unjust political structures, as Marxism holds. Christian doctrine opposes the error according to which evil is only caused by the subconscious (Freud) or by psychological conditioning (behaviourism). Evil is not a necessary part of evolutionary processes either, as followers of Darwin and Huxley sustain. Evil does not derive from genetic randomness or determination, as some atheists such as Dawkins sustain. It does not derive from the oppression of women, as the feminists propose: this is a confusion of cause and effect.

Genesis chapter 3 provides the biblical description of original sin. In any discussion of the ecological threat, both original and personal sin must be fully taken into account. The episode of the flood is also an indication of a deviation from the divine will. It begins with the affirmation that all flesh (man and animals) had contaminated the earth through increasing wickedness, filling it with violence and corruption and leading to its ruin. Sin is understood as 'violence' (*hamas*) or 'lawlessness' which corrupts God's work: 'In the eyes of God, the earth was corrupt and full of lawlessness. When God saw how corrupt the earth had become, since all mortals led depraved lives on earth...' (Gn 6:11–12). At the same time, however, God's faithfulness to the covenant is again proclaimed (Gn 8:21–9:7), and the animals that were with man are also included in the divine covenant (Gn 9:8ff).

The prophet Isaiah affirms: 'The earth is polluted because of its inhabitants, who have transgressed laws, violated statutes, broken the ancient covenant' (Is 24:5). The prophet Hosea echoes this same sentiment: 'Therefore the land mourns, and everything that dwells in it languishes: the beasts of the field, the birds of the air, and even the fish of the sea perish' (Ho 4:3).

The ecological threat comes from humanity and not from the untamed forces of a wild and indomitable nature. Man, the sinner, threatens the balance and harmony of the world, which becomes ambiguous and capricious. As the prophet Jeremiah affirms: 'For the wickedness of those who dwell in it beasts and birds disappear, because they say, "God does not see our ways"' (Jr 12:4b). Original sin, however, does not remove man's intellect and will, so he always remains free in his actions. Furthermore, God does not hold back on His promise, which involves a constantly renewed possibility of saving the balance and harmony of the world.

St. Paul affirms that 'creation was made subject to futility, not of its own accord but because of the one who subjected it' (Rm 8:20). Here the question arises as to whether the nature of the fallenness of the cosmos after original sin was essential or *per accidens*. God allowed the cosmic repercussions of original sin in order to show us the redemption of the cosmos. However what are the collateral effects of original sin: that viruses and bacteria are dangerous and cause illness, and that earthquakes, volcanoes, tsunamis and asteroids that strike the earth are destructive and devastating?

We can identify various interpretations. One of these proposes the idea that before original sin, human beings had a special relational capacity of managing the cosmos, even in its most difficult and negative manifestations, such as ferocious animals. The second possibility is that before the Fall, there were no difficult aspects or obstacles. Here, the distinction must be made between physical and moral evil. Before the fall of man into sin, there was no moral evil on the earth.

How should the phrase 'because of the one who subjected it' be understood? Non–human creation is subject to fallenness without its own will or decision. Fallenness is the cosmic repercussion of original sin. In Genesis 3:16f, creation is subjected to fallenness as an effect of Adam's sin. It must also be remembered that the Church firmly teaches that the devil and demons do exist.[63] Paul VI reaffirmed that the devil 'exists and that he is still at work with his treacherous cunning; he is the hidden enemy who sows errors and misfortunes in human history.' The devil, in fact, can exert a malign and seductive influence 'on individuals as well as on communities, entire societies or events.'[64] The devil is the 'cosmic vandal.'

In any event, disorder entered the cosmos after and as a consequence of original sin:

> When Anne Frank died at sixteen years of age in the Bergen–Belsen concentration camp in 1945, we knew who to accuse. When a sick girl of the same age dies in the hospital of an absurd and incurable disease, we don't know who to accuse ... War is first of all a form of the natural violence which manifests itself in earthquakes or biological deviations. The Bible leaves no doubts: a tragedy has disrupted man's relation to his Creator since the beginning.[65]

4.3.7 Redemption

In order to appreciate the value of redemption, evil must be taken seriously. We must remember that Christian ecology is not something made only by human hands. It is the work of God and His grace, but with human cooperation. The cross of Christ represents victory over the disorder and iniquity inherent to man's poor use of nature. Only by applying the power of Christ Crucified and Risen, through the Church, can we reestablish peace with nature. We must avoid the error of thinking about the ecological issue only in terms of social and political solutions. The mystery of evil has been conquered not only through human action and work, but by the grace of God in Christ Jesus, in the sacraments, which make use of elements deriving from nature (bread, wine, oil, water). Christ redeems the cosmos through the Church, through the human beings who are His ministers, through the human person, who is a microcosm of the universe; Christ brings peace to the cosmos.

Through the blood of His Cross and through His Resurrection, Christ has restored to creation its original order. Henceforth the whole world, with man at its centre, has been snatched from the slavery of death and corruption (Rm 8:21) and in a certain sense has been created anew (Rev 21:5); it now exists no longer for death but for life, for new life in Christ. Thanks to his union with Christ, man rediscovers his proper place in the world. In Christ he experiences anew that original harmony which existed between Creator, creation and man before man succumbed to the effects of sin. In Christ, man re–reads his original call to subdue the earth, which is the continuation of God's work of creation rather than the unbridled exploitation of the cosmos.[66]

The announcement of redemption, which offers a hope of salvation beyond a merely earthly future, cannot avoid having consequences for ecological ethics, which, among other things, invites everyone to take responsibility for the future of this world. Faith in Jesus Christ does not excuse man from concern for the world, it does not resolve the technical problems of the world for him, but it does form him to an attitude of service to the world, service done with gratitude and which does not halt in the face of any defeat. This world, in fact, was given to him not just as a gift but also as a responsibility: human beings must work the land and preserve it. What man does in this world certainly has an impact also in the other world, beyond history, as the 'management parables' of the New Testament demonstrate (cf. Mt 25:14ff; Mk 12:1ff).[67] Saint Maximus the Confessor affirmed:

The mystery of the Incarnation of the Word contains in itself all the secrets and enigmas of the Holy Scriptures and the hidden meaning of all visible creatures, but he who knows the mystery of the Cross and of the empty tomb knows the essential reasons of all things, and he who is initiated into the arcane power of the resurrection knows the aim for which God created all things *in principio*.[68]

4.3.8 The Church and the cosmos

The relationship between the Church and the cosmos appears primarily in Saint Paul's letters to the Ephesians and the Colossians. According to the Letter to the Ephesians , God 'put all things beneath His [Jesus Christ's] feet and gave Him as head over all things to the church, which is His body, the fullness of the one who fills all things in every way' (Ep 1:22f.). There is a double lordship of Christ, both of which are connected to His kingship. The first is with respect to the universe, 'all things,' of which Christ is the Head in the sense of Lord (Ep 4:10; Ph 2:9–11). The second concerns the Church, of which Christ is the Head as its sustenance and life force in the sense of grace (Ep 1:22f).

It is important to point out that, according to the Letter to the Ephesians, the cosmos is never presented as 'the Body of Christ.' Only the Church is the Body of Christ, and this closes the door to cosmocentrism and the organismic idea. Through the Church, Christ brings to completion the fullness of the world. The Church must bring man and the world to salvation: '"Go into the whole world and proclaim the Gospel to every creature..."' (Mk 16:15). In the Letter to the Colossians, the following is read:

He [Christ] is the beginning, the firstborn from the dead, that in all things he himself might be preeminent ... For in him all the fullness was pleased to dwell, and through him to reconcile all things for him, making peace by the blood of his cross [through him], whether those on earth or those in heaven (Col 1:18ff).

The reconciliation of the universe occurs through the Sacrifice of Christ on the Cross, which is applied through the Church particularly in the Sacrifice of the Mass. Through the Church, the glorified Lord unites the cosmos to Himself together with redeemed humanity in an ever more profound and efficacious way. The Church is the organ through which the unification of the universe in Christ, provided for in the eternal plan for the world, is actualized through history. The most ancient liturgical rites envisage all of creation included in the Eucharist of the Church, in the Sacrifice and sacrament of the Mass. The Eucharist is the true source of the reasons for a Christian ecology. Only one who is united with the Eucharist perceives creation as a gift from God made in Christ and in the power of the Holy Spirit; only one united in with Eucharist understands how all of creation, a community of co–creatures, is in relation with Christ, the firstborn of all creatures. Only one united with the Eucharist knows how to expectantly await, *donec veniat* (cf. 1 Co 11:26), a new heaven and a new earth, when God will be all in all (cf. 1 Co 15:28).

The cosmic dimension of the Church was already established by Origen. He identified the Church as the adornment of the world, and is said to be the light of the world. 'Now, the adornment of the world is the Church, Christ being her adornment, who is the first light of the world.'[69] Christ applies the fruits of redemption to the cosmos through the Church. The Church is the only sacrament of salvation. The Church is the efficacious centre of sacredness in the universe. While the act of redemption is complete in itself, its application to the cosmos must be brought to completion. In the East, the cosmos is considered the temple in which humanity carries out its priestly role in a theocentric perspective. In the West, on the other hand, the cosmos is understood as the home in which man is the administrator and caretaker in an anthropocentric perspective. This Western perspective is limited, because the cosmos is not renewed merely through human work.

4.3.9 The Holy Spirit and creation

The Spirit of God, which 'fills the world' (Ws 1:7), has not ceased sowing plentiful seeds of truth, love and life in the hearts of the men and women of our time.[70] These seeds have produced the fruits of progress, humanization and civilization, which constitute authentic signs of hope for humanity on its journey. The 'greater awareness of our responsibility for the environment' represents a sign of hope.[71] Today, as part of a reaction to the reckless exploitation of natural resources which has often accompanied industrial development, humanity is rediscovering the meaning and value of the environment as a hospitable dwelling (*oúkos*) within which it is called to live. The threats which weigh on humanity's future, due to the lack of respect for the balance of the ecosystem, push people of culture and science and the competent authorities to study and enact various projects and measures. They aim not just to limit or repair the damage which has been caused thus far, but above all to delineate a harmonious form of societal development which respects and values the natural environment.

This active sense of responsibility for the environment must also stir Christians to rediscover the profound meaning of the creative plan revealed by the Bible. God wished to entrust man with the task of filling the earth and being its lord in His name, almost as His representative (cf. Gn 1:28), prolonging and in a certain way bringing to completion His creative work itself.[72]

God created the world for His Son in the Holy Spirit: 'a mighty wind [the Holy Spirit] swept over the waters' (Gn 1:2). The Wisdom books contain allusions to the presence of the Holy Spirit in creation. The act of creation is a Trinitarian act. Therefore, everything that exists was created by God the Father through God the Son, in God the Holy Spirit. If the Holy Spirit is spread throughout all of creation, then the Spirit creates the community of all created things with God and with one another, actualizing that communion of creation in which all created things, each in its own way, communicate with one another and with God. The presence of the Holy Spirit in the world creates the general harmony and symmetry found in a hierarchical way in all natural relationships. The Holy Spirit again offers that beauty which was disfigured by sin, and helps us to find peace with nature:

> We know that all creation is groaning in labour pains even until now; and not only that, but we ourselves, who have the first fruits of the Spirit, we also groan within ourselves as we wait for adoption, the redemption of our bodies. (Rm 8:22–23)

We are distinct from the rest of creation because we already possess the first fruits of the Holy Spirit. Sin alienated us not just from God, not just from one another and from our true selves, but also from nature. It even alienated nature from its true purpose. But nature, creation itself, can 'hope' to be freed from the slavery of corruption. This liberation is carried out by the Holy Spirit which acts in Christians. The work of the Holy Spirit will continue until eschatological fulfilment.

4.3.10 Eschatological perspectives

The more the decision for the future is subjected to dark prophesies of a threatening collapse of the fundamental conditions for our lives, the more new problems emerge: is humanity perhaps definitively faced with its ultimate end? Must there always be a history that continues forever and provides us with some sort of guarantee? Or must there rather be, sooner or later, an end to history? And what will happen then? Let us now consider the most important forms of 'world fulfilment' theology. Only God could create the universe; only He can recreate it and bring it to fulfilment. Retrogressive and progressive utopianism must be avoided. In optimism, there is a total continuity between the current world and the afterlife, whereas in pessimism there is the total destruction of the current universe. In the Catholic eschatological perspective, instead, the second coming of Christ can be seen as the event by which God physically comes to reside in the perfected universe which brings to fulfilment the original plan of creation. When Christ comes in His glory, He will 'recapitulate' all of creation in a moment of definitive and eschatological harmony.[73]

a) Traditional apocalyptics

Apocalyptics sustains that, in accordance with a preor-dained plan, sooner or later God will end the battle between Himself and the sinful world and will establish His total Lordship. For this reason, attention is placed especially on those events and historical developments which can be interpreted as signs of the imminent end. Incumbent self–annihilation at the hands of atomic weapons is, for some, the clear sign that we find ourselves harshly faced with a 'threshold'! Some fundamentalist Protestants sustain this position. The New Testament itself, however, has insistently warned against any attempt to calculate the end in terms of time.

b) Teleological eschatology

Traditional theology understands eschatology as a 'doctrine of final things'. On the basis of Scripture and Tradition, it attempts to describe 'the future events, presented as least tendentially in a realistic form, which will occur at the end of history.'[74] Sometimes there is more interest in a certain 'physics of final things' than in the true meaning of the biblical affirmations. Final fullness, in fact, has already begun in a mysterious way through man's reception of the gift of grace. The time of the Church is to be understood as the intermediate, definitive time, which will sooner or later pass into the fulfilment of the kingdom of God, which all will experience. The special value of the *eschaton* and the distinction between the 'already' and the 'not yet' must be remembered. The connection between the end of the world and the second coming of Christ must also be affirmed.

c) Historic–salvific–realistic eschatology

This attempts to understand the whole course of history in a unitary perspective: time and history have been moving, since the beginning, in the direction of the central event of the death and resurrection of Jesus Christ. What happened in Jesus Christ in a representative way will be extended by the *parousia* to the entire universe and be definitively actualized: it will be the new creation. The history of salvation and the history of the world are like the image of a double concentric circle: the interior circle of the history of salvation radiates ever closer to the exterior circle of the history of the world, actively working toward general salvation. The end involves the liberation and transformation of the entire universe. This position is held and represented by O. Cullmann.[75]

d) Futuristic eschatology

This seeks, as much as possible, to closely relate the secular understanding of the future and the Christian eschatological hope for it. In its extreme form, it defends the conviction that the eschatological salvation of the God Who comes can be actualized only together with the humanization of the earth; in fact, they coincide. This is the position, for example, of some expressions of liberation theology. It is a very questionable position insofar as it is too horizontalist and immanentist. Along this same line is Pierre Teilhard de Chardin's view, which sees human and divine action at work in a general evolutionary process toward the manifestation of the *eschaton*. The *eschaton* seems to be a necessary product of this evolution, in a notion that can also be found in the New Age Movement. Biological evolution cannot be identified with the history of salvation in the same way. Jürgen Moltmann attempts to envision the 'future' of God and the future of human history together; nonetheless, in his opinion, the two moments are not identical.

e) Eschatology reduced to existential ethics

This is also an erroneous view because it is marked by a reduction. The classic example is given by Rudolf

Bultmann's position, which holds that the end of the world and the *parousia* did not come about as was expected, that history has continued to progress, and that, according to all reasonable people, it will continue to progress.[76] According to Bultmann, the *eschaton* consists in the fact that Christ freed us to act responsibly in history. The eschatological hour of the *parousia* does not occur at the end of history, but is rather that event within history by means of which Christ opened for us the possibility of true decisions. The meaning of history is in the present, and when the present is understood by Christian faith as the eschatological present, it is then that the meaning of history is actualized. In this position, the doctrine of God's transcendence is lacking. In existentialism, every moment has the same weight without there being privileged moments, and this resembles the error of communism which also excludes social privileges.

f) Verticalist eschatology

In recent Protestant theology, the apocalyptic idea of time and history has been abandoned together with the apocalyptic idea of space. Karl Barth, for example, shows no interest in an end–of–history eschatology. The 'eternal moment' does not coincide with a historical moment, but can be found as part of every historical moment as its 'transcendent meaning.' The 'end' is near at every moment. Barth emphasizes, in his interpretation of Romans 13:11ff, that eternity is the overcoming of all time.[77] The end announced in the New Testament is not a temporal event. The 'end' expresses God's sovereign transcendence over the nothingness of creatures. In this position, God's immanence is lacking. Catholic theology, instead, teaches that the *parousia* impacts the history of man and the cosmos.

g) Biblical vision of cosmic reconciliation

This beautiful biblical vision of cosmic reconciliation comes from the Old Testament. This comes from a careful reading of the passage Isaiah 11:6–9, which contains two meanings. The first is a looking back to the paradisiacal state, before original sin. The second meaning is a looking forward to eschatological reconciliation, indicated by the future tenses of the verbs:

> Then the wolf shall be a guest of the lamb, and the leopard shall lie down with the kid; the calf and the young lion shall browse together, with a little child to guide them. The cow and the bear shall be neighbours, together their young shall rest; the lion shall eat hay like the ox. The baby shall play by the cobra's den, and the child lay his hand on the adder's lair. There shall be no harm or ruin on all my holy mountain; for the earth shall be filled with knowledge of the Lord, as water covers the sea. (cf. Is 65:25)

h) The beginning of the eschata in death

The dialectics of time and eternity, as developed by recent Protestant theology and especially by dialectical theology, risks undervaluing history and human action in history. To obviate this danger, Lohfink uses the term *aevum*, which was used by the Scholastics and especially Saint Thomas Aquinas. *Tempus* is the continual flowing of the moments of history: it has a beginning and an end. *Aeternitas* is the mode of being of God, in which He possesses all of His being in a present which includes all things, *tota simul et perfecta possessio*, without beginning or end, according to the Boethian interpretation.[78] Between *tempus* and *aeternitas* is the *aevum*, which has a beginning but no end.

Furthermore, G. Lohfink uses the concept of transfigured time rather than *aevum*.[79] Transfigured time is distinguished from historical time by the fact that in it there is no longer any before or after: it includes, in a single present, all of human existence. Transfigured time is distinguished also from eternity in that man's new mode of being is 'constituted by time.' Transfigured time reflects the *aeternitas*; it is not merely the result of the continual maturation of human existence in its completion in the afterlife, but also the process itself by which the history of man is brought before God.[80] Corporality, being part of the world and all social communication are forever 'inscribed' in the subject, and in death are assumed into subject's definitive mode of being as once they were gathered up by him in the course of time. This implicates a certain fulfilment. Thus, Lohfink considers the *eschata* to begin upon the death of individual human beings. The material universe is a means of the human history of freedom in such a fundamental way that it must be included in man's fulfilment. The material universe is thereby transfigured. The relationship between *kairos* and *cronos* is also relevant to this topic.[81]

i) The Holy Eucharist, pledge and first fruits

In the Eucharist, everything expresses the faithful waiting 'in joyful hope for the coming of our Saviour, Jesus Christ.'[82] According to Pope John Paul II, one who is nourished by Christ in the Eucharist does not have to wait for the afterlife in order to receive eternal life: it is already had on earth, as the first fruits of the future fullness of man in his totality.[83] In the Eucharist, in fact, we also receive the guarantee of bodily resurrection at the end of the world: 'Whoever eats my flesh and drinks my blood has eternal life, and I will raise him on the last day' (Jn 6:54). This guarantee of future resurrection derives from the fact that the flesh of the Son of Man, given as food, is His Body in the glorious state of resurrection. Through the Eucharist, the 'secret' of the resurrection is assimilated. Therefore, rightly, Saint Ignatius of Antioch defined the Eucharist as the 'medicine of immortality, and the antidote to prevent us from dying.'[84]

The rightful concerns about the ecological conditions in many parts of the created world find comfort in the perspective of Christian hope, which commits us to work responsibly to safeguard creation. In the relationship between the Eucharist and the cosmos, in fact, we discover the unity of God's design and are led to grasp the profound relationship between creation and the 'New Creation,' inaugurated by the resurrection of Christ, the New Adam. We participate in this new creation already by way of our Baptism (cf. Col 2:12f), and so, nurtured by the Eucharist, our Christian life opens us to the prospect of a new world, of a new heaven and a new earth, where the new Jerusalem descends from heaven, from God, 'prepared as a bride adorned for her husband' (Rev 21:2).[85] The eschatological tension evoked by the Eucharist expresses and reinforces the communion with the heavenly Church. A significant consequence of this eschatological tension inherent in the Eucharist is also the fact that it encourages us on our human journey through history, instilling a seed of living hope in the daily dedication of each individual to his or her given tasks.[86] If, in fact, the Christian vision leads to looking toward 'a new heaven' and 'a new earth' (cf. Rev

21:1), this does not weaken, but rather encourages our sense of responsibility toward the present earth.[87]

The theology of the Orthodox Churches also applies the eschatological dimension of the Eucharist to the theology of the environment. This position is exemplified, in fact, by Zizioulas.[88] The Eucharist, in its most intimate nature, contains an eschatological dimension which, for as much as it penetrates history, never fully transforms into history and thereby transcends history. The Eucharist will open the road not to the dream of the moral perfection of the world (according to an evolutionary framework), but to the need for the radical exercise and experience of the *kenosis* and the cross, the only way to live the victory of the resurrection in the world until the end of time. The Eucharist, nonetheless, will simultaneously offer the world a taste of the eschatological reality, which penetrates history through the Eucharistic assembly and makes our divinization in space and time possible.

Notes

1   See p. 114 above.
2   Cf. Cardinal P. Poupard, *Address for the presentation of the Jubilee of Scientists* (February 28, 2000).
3   Cf. S. L. Jaki, *Science and Creation* (Edinburgh: Scottish Academic Press, 1986).
4   Vatican II, *Ad Gentes Divinitus*, 8.
5   Cf. S. L. Jaki, *The Savior of Science* (Edinburgh: Scottish Academic Press, 1990), pp. 43–45.
6   Cf. P. Haffner, *The Mystery of Reason* (Leominster: Gracewing, 2001), pp. 117–124.
7   S. L. Jaki, *The Road of Science and the Ways to God* (Edinburgh: Scottish Acdemic Press, 1978), p. 298.

8   Cf. L. White, 'The historical roots of our ecological crisis,' in *Science* 155 (1967), pp. 1203–1207.

9   This topic will be addressed in section 4.3.6 below.

10   C. Améry, *Das Ende der Vorsehung. Die gnadenlosen Folgen des Christentums* (Hamburg 1972).

11   J. Passmore, *Man's Responsibility for Nature* (London 1974, ²1980).

12   A. Næss 'The Shallow and the Deep, Long-Range Ecology Movement' in *Inquiry* 16 (1973), pp. 95-100.

13   Cf. F. Capra, *The Turning Point. Science, Society and the Rising Culture* (New York, 1982).

14   On the relationship between environmentalism and feminism, cf. J. Cheney, 'Ecofeminism and Deep Ecology,' in *Environmental Ethics* 9 (1987), pp.115–146; I. Diamond, G. F. Orenstein, *Reweaving the World. The Emergence of Ecofeminism* (San Francisco, 1990); R. Whelan, J. Kirwan, P. Haffner, *The Cross and the Rain Forest* (Grand Rapids, MI: W. B. Eerdmans, 1996), p. 128.

15   The term was coined by British psychologist Richard D. Ryder in 1973 to denote a prejudice based on physical differences.

16   W. Kasper, 'La sfida ecologica alla teologia' in A. Caprioli e L. Vaccaro, *Questione ecologica e coscienza cristiana* (Brescia: Morcelliana, 1988), p. 134: 'In effetti la fede biblica della creazione fa parte dei presupposti spirituali delle moderne scienze naturali e dello sviluppo tecnologico che esse hanno reso possibile, poiché è stata la distinzione biblica fra Creatore e creatura a smitologizzare e dedivinizzare il mondo come creazione di Dio. Anche un mondo razionale è creato da Dio razionale. Il movimento ecologico si è allontanato dalla visione giudaico–cristiana. Nel distacco dell'antropocentrismo biblico, il movimento ecologico sostiene non solo un'unità e un'osmosi quasi mistica tra uomo e mondo, ma anche una specie di definizione panteistica del rapporto fra Dio e il mondo. Questa visione

di alcuni ecologi presenta una nuova sfida alla teologia.'

[17] Cf. E. Bardulla, 'I cristiani di fronte alla questione ambientale' in A. Caprioli e L. Vaccaro, *Questione ecologica e coscienza cristiana* (Brescia: Morcelliana, 1988), p. 119. See also J. D. Hughes, *Ecology in Ancient Civilisations* (Albuquerque: University of New Mexico Press, 1975).

[18] See M. Weber, *The Protestant ethic and the spirit of capitalism* (Mineola, NY: Dover Publications, 2003).

[19] Here the terms 'person' and 'nature' are used in their concrete, and not philosophical, meanings.

[20] International Theological Commission, *Communion and Stewardship: Human Persons Created in the Image of God* (23 July 2004), 74.

[21] Cf. *ibid.*, 78.

[22] Cf. P. Haffner, *The Mystery of Reason* (Leominster: Gracewing, 2001), pp. 12–19.

[23] Cf. German Bishops' Conference, *Safeguarding the future of creation* (1998), 66.

[24] Cf. G. Greshake, 'La creazione come autorivelazione e dono di sé da parte di Dio' in A. Caprioli e L. Vaccaro, *Questione ecologica e coscienza cristiana*, p. 128.

[25] St Anselm of Canterbury, *Monologion*, 33.

[26] St Bonaventure, *In II sententiarum* 16, 1, 1 .

[27] *Ontologism* is the idea that the human intellect has as its proper object the knowledge of God, that this knowledge is immediate and intuitive, and that all other knowledge must be built on this base. Malebranche developed this theory under the influence of Platonic and Cartesian philosophies. Vincenzo Gioberti developed his notion of Ontologism, and its fundamental principles are present in Rosmini's philosophy, although there have been many attempts to defend him against this accusation.

[28] Theodoret of Cyrrhus, *Discorsi sulla provvidenza divina*, 10: 'L'incarnazione del nostro Salvatore rappresenta il più alto

compimento della sollecitudine divina per gli uomini. Infatti né il cielo né la terra né il mare né l'aria né il sole né la luna né gli astri né tutto l'universo visibile e invisibile, creato dalla sua sola parola o piuttosto portato alla luce dalla sua parola conformemente alla sua volontà, indicano la sua incommensurabile bontà quanto il fatto che il Figlio Unigenito di Dio, Colui che sussisteva in natura di Dio, riflesso della sua gloria, impronta della sua sostanza, che era in principio, era presso Dio ed era Dio, attraverso cui sono state fatte tutte le cose, dopo aver assunto la natura di servo, apparve in forma di uomo, per la sua figura umana fu considerato come uomo, fu visto sulla terra, con gli uomini ebbe rapporti, si caricò delle nostre infermità e prese su di sé le nostre malattie.'

29 German Bishops' Conference, *The Future of Creation — the future of humanity* (1980), II,1 and III,1.

30 Cf. St Irenaeus, *Adversus Haereses*, Book V, chapter 4, 1, where he says that not only the soul but also the body is important. See also International Theological Commission, *Communion and Stewardship: Human Persons Created in the Image of God*, 27: 'This truth has not always received the attention it deserves. Present–day theology is striving to overcome the influence of dualistic anthropologies that locate the *imago Dei* exclusively with reference to the spiritual aspect of human nature. Partly under the influence first of Platonic and later of Cartesian dualistic anthropologies, Christian theology itself tended to identify the *imago Dei* in human beings with what is the most specific characteristic of human nature, viz., mind or spirit. The recovery both of elements of biblical anthropology and of aspects of the Thomistic synthesis has contributed to the effort in important ways.'

31 Cf. International Theological Commission, *Communion and Stewardship: Human Persons Created in the Image of God*, 9.

32 *Ibid.*, 76.

33 Cf. Vatican II, *Gaudium et spes*, 24 where it is affirmed that man 'is the only creature on earth which God willed for itself.'

34 Cf. A. Bonora, 'L'uomo coltivatore e custode del suo mondo in Gen 1–11' in A. Caprioli e L. Vaccaro, *Questione ecologica e coscienza cristiana* (Brescia: Morcelliana, 1988), p. 161. The Hebrew expression is כְּבַשׁ.

35 *Ibid*. Here, the Hebrew term being discussed is רָדָה.

36 N. Lohfink, *Le nostre grandi parole* (Brescia: 1986), pp. 192–193: 'questa benedizione non legittima affatto la distruzione di intere famiglie di animali nei diversi continenti, di esseri marini, delle innumerevoli specie di volatili e di insetti, in nome della trasformazione della superficie terrestre da parte dell'uomo... Questa benedizione significa il contrario.'

37 Cf. St Thomas Aquinas, *Summa Theologiae*, Iᵃ–IIae, q. 2, a. 8 arg. 2: 'Praeterea, ultimus finis cuiuslibet rei est in suo perfecto, unde pars est propter totum, sicut propter finem. Sed tota universitas creaturarum, quae dicitur maior mundus, comparatur ad hominem, qui in VIII Physic. dicitur minor mundus, sicut perfectum ad imperfectum. Ergo beatitudo hominis consistit in tota universitate creaturarum.'

38 Cf. A. Bonora, 'L'uomo coltivatore e custode del suo mondo in Gen 1–11' in A. Caprioli e L. Vaccaro, *Questione ecologica e coscienza cristiana*, p. 162.

39 Cf. German Bishops' Conference, *Safeguarding the future of creation* (1998), 68.

40 Cf. A. Bonora, 'L'uomo coltivatore e custode del suo mondo in Gen 1–11' in A. Caprioli e L. Vaccaro, *Questione ecologica e coscienza cristiana*, p. 163.

41 St Thomas Aquinas, *Summa Theologiae* I, 103, 6.

42 German Bishops' Conference, *Safeguarding the future of creation* (1998), 64–65.

43 See pp. 116–117 above.

44   Cf. Pope John Paul II, *Address to the Pontifical Academy of Sciences* (23 October 1982), in *IG* 5/3 (1982). Cf. *CCC* 2418.

45   Cf. M. Damien, *Gli animali, l'uomo e Dio* (Casale Monferrato: Piemme, 1987), p. 32.

46   *Ibid.*, p.166: 'Cristo è morto anche per i cani.'

47   St Augustine, *City of God*, Book XX, chapter 16 in P. Schaff (ed.), *Nicene and Post–Nicene Fathers*, First Series, Volume 2, *St. Augustine's City of God and Christian Doctrine* (Peabody: Hendrickson, ²1995), p. 435.

48   St Thomas Aquinas, *Summa Theologiae Supplement*, q.91, a.5.

49   Cf. J. Galot, 'Il destino finale dell'universo' in *La Civiltà Cattolica* 152/4 (November 3, 2001), pp. 213–225; G. Cavalcoli, 'La dimensione escatologica del tempo secondo la rivelazione cristiana' in *Sacra Dottrina* 44/1 (1999), pp. 5–46.

50   International Theological Commission, *Communion and Stewardship: Human Persons Created in the Image of God*, 80.

51   Cf. P. Haffner, *Mystery of Creation* (Leominster: Gracewing, 1995), p. 214. Cf. also International Theological Commission, *Communion and Stewardship: Human Persons Created in the Image of God*, 76: 'Not only human beings, but the whole of visible creation, are called to participate in the divine life.'

52   Cf. S. L. Jaki, *Science and Creation* (Edinburgh: Scottish Academic Press, 1986). Cf. also P. Haffner, *Mystery of Creation*, pp. 135–136.

53   Cf. P. Haffner, *Creation and Scientific Creativity: A Study in the Thought of S. L. Jaki* (Front Royal: Christendom Press, 1991), p. 108.

54   Vatican II, *Lumen Gentium*, 2 and cf. Col 1:15.

55   Cf. G. Biffi, 'Fine dell'Incarnazione e Primato di Cristo' in *La Scuola Cattolica* 80 (1960), p. 251.

56   Cf. Pope John Paul II, *General Audience* (14 February 2001), 1.

57   St Irenaeus, *Adversus Haereses*, Book III, cap. XXI, 9. In the expression 'all things' — Irenaeus affirms — man is included, touched by the mystery of the Incarnation, such

that the Son of God 'took up man into Himself, the invisible becoming visible, the incomprehensible being made comprehensible, the impassible becoming capable of suffering, and the Word being made man, thus summing up all things in Himself: so that as in super–celestial, spiritual, and invisible things, the Word of God is supreme, so also in things visible and corporeal He might possess the supremacy, and, taking to Himself the pre–eminence, as well as constituting Himself Head of the Church, He might draw all things to Himself at the proper time.' (*Adversus Haereses* III, XVI, 6).

[58] Cf. Pope John Paul II, *General Audience* (14 February 2001), 2.

[59] Cf. *ibid.*, 3.

[60] St Irenaeus, *Adversus Haereses*, Book V, c. XIV, 1; cf. Book V, c. XIV, 2.

[61] See Pope John Paul II, *Address at the Liturgy of the Word in Zamosc* (12 June 1999), 3.

[62] Cf. Pope John Paul II, *General Audience* (14 February 2001), 4.

[63] Cf. R. Faricy, *Wind and Sea obey Him* (London: SCM, 1982), pp. 41-42.

[64] Pope Paul VI, *Discourse at General Audience* (15 November 1972). Cf. *CCC* 395.

[65] M. Damien, *Gli animali, l'uomo e Dio* (Casale Monferrato: Piemme, 1987), p.131: 'Quando Anna Frank muore a sedici anni nel campo di concentramento di Bergen–Belsen nel 1945, sappiamo chi accusare. Quando una malata della stessa età muore in ospedale di una malattia assurda e interminabile, non si sa più chi accusare... La guerra è prima di tutto una forma della violenza naturale che si manifesta nei sismi o nelle deviazioni biologiche. La Bibbia non lascia dubbi: un dramma ha sconvolto fin dalla comparsa dell'uomo i suoi rapporti con il Creatore.'

[66] See Pope John Paul II, *Address at the Liturgy of the Word in Zamosc* (12 June 1999), 4.

67   Cf. Common Declaration of the Catholic and Evangelical Churches in Germany, *Feeling responsibility for creation* (14 May 1985), 68.

68   St Maximus the Confessor, *Capitoli teologici*, I, 66 in *PG* 90, 1108 : 'Il mistero dell'Incarnazione del Verbo contiene in sé tutti i segreti e gli enigmi delle Sante Scritture e il senso nascosto di tutte le creature visibili, ma colui che conosce il mistero della Croce e della tomba vuota conosce le ragioni essenziali di tutte le cose, e chi è iniziato all'arcana potenza della risurrezione conosce lo scopo per cui Dio ha creato *in principio* tutte le cose.'

69   Origen, *Commentary on the Gospel of John*, Book VI, n.38 in *PG* 14, 301–302.

70   Cf. Vatican II, *Gaudium et spes*, 11.

71   Pope John Paul II, Apostolic Letter *Tertio millennio adveniente*, 46.

72   Cf. Pope John Paul II, *Discourse at General Audience* (18 November 1998), 1, 3.

73   International Theological Commission, *Communion and Stewardship: Human Persons Created in the Image of God*, 75, 79.

74   G. Greshake & G. Lohfink, *Naherwertung, Auferstehung, Unsterblichkeit. Untersuchungen zur christlichen Eschatologie* (Freiburg: Herder, 1982), p. 12.

75   Cf. O. Cullmann, *Christ and Time: The Primitive Christian Conception of Time and History* (Philadelphia: Westminster Press, 1950).

76   Cf. R. Bultmann, *Geschichte und Eschatologie im Neuen Testament*, in *Glauben und Verstehen III* (Tübingen: 1960), pp. 91–106.

77   Cf. K. Barth, *Der Römerbrief* (Zürich: Evangelischer Verlag, 1954).

78   Boethius, *De consolatione philosophiae*, Book V, 6, in *PL* 63, 858–859.

79   G. Greshake & G. Lohfink, *Naherwertung, Auferstehung, Unsterb-*

lichkeit. Untersuchungen zur christlichen Eschatologie, p. 67f.

80 Cf. also J. H. Newman, *The Dream of Gerontius* where the following can be found:

Nor touch, nor taste, nor hearing hast thou now;
Thou livest in a world of signs and types,
The presentations of most holy truths,
Living and strong, which now encompass thee.
A disembodied soul, thou hast by right
No converse with aught else beside thyself;
But, lest so stern a solitude should load
And break thy being, in mercy are vouchsafed
Some lower measures of perception,
Which seem to thee, as though through channels brought,
Through ear, or nerves, or palate, which are gone.
And thou art wrapped and swathed around in dreams,
Dreams that are true, yet enigmatical;
For the belongings of thy present state,
Save through such symbols, come not home to thee.
And thus thou tell'st of space, and time, and size,
Of fragrant, solid, bitter, musical,
Of fire, and of refreshment after fire;
As (let me use similitude of earth,
To aid thee in the knowledge thou dost ask) –
As ice which blisters may be said to burn.

81 Cf. P. Haffner, 'The mission of Christ in time: past, present, future' in *Studia Missionalia* 52(2003), pp. 229–244.

82 Roman Missal, *Embolism after the Our Father*.

83 Pope John Paul II, Encyclical Letter *Ecclesia de Eucharistia* (2003), 18.

84 St Ignatius of Antioch, *Epistle to the Ephesians*, XX in *Ante-Nicene Fathers, Volume 1, The Apostolic Fathers, Justin Martyr, Irenaeus* (Peabody: Hendrickson, ²1995), p. 58.

85 Cf. Pope Benedict XVI, Apostolic Exhortation *Sacramentum Caritatis*, 92.

86 Cf. Pope John Paul II, Encyclical Letter *Ecclesia de Eucharistia*,

19, 20.

[87]   Cf. Vatican II, *Gaudium et Spes,* 39.

[88]   Cf. I. Zizioulas, *Il creato come eucaristia. Approccio teologico al problema dell'ecologia* (Magnano: Qiqajon, 1994).

5

Morality and Spirituality

Threats and scourges and destruction hang over us, Lord, because of the multitude of our transgressions; for we have sinned and transgressed and gone far from you, and we are affected and afflicted by dire perils; but deliver us, Lord from dangers that beset us, and keep the whole fabric of the earth unharmed, granting equable breaths of wind and ever– flowing springs of water for our safe–keeping and salvation, O Lover of humankind.

Gerasimos of the Skete of Saint Anne, Creation Vespers

5.1 Ecology and Morality

The New Age Movement is known for its refutation of the notion of sin — considered inescapably dogmatic — and the substitution of illness in its stead. The New Age Movement does not deny that inappropriate behaviours exist in the world: just think of the horror that environmentally harmful actions instil in its adherents. However, it ascribes these actions to physical or psychological limitations which can be likened to illness or forms of 'addiction' which can be overcome through various forms of therapy and recovery which are widely available within the New Age Movement itself. Social evils can be overcome as well through a general awareness, which will automatically resolve the problems of the world, and it is here where the core of New Age politics can be found.

The Christian vision, instead, affirms that original and actual sin exist. Sin is an opposition to reason, truth, and right conscience; it is a transgression against sincere love of God and neighbour, due to a perverse attachment to certain goods. It harms human nature and threatens human solidarity. It has been defined as 'a word, act or desire contrary to the eternal law.'[1]

As Saint Paul affirmed: 'where sin increased, grace overflowed all the more.' Grace, however, in order to carry out its task, must unveil sin in order to convert our hearts and grant us 'justification for eternal life, through Jesus Christ our Lord' (Rm 5:20–21). As a doctor who examines a sore before treating it, God casts the light of life upon sin with His Word and His Spirit.[2] Looking into the Face of God, man can bring light to the face of the earth and ensure, through ethical commitment, a welcoming and healthy environment for current and future generations.[3]

Unfortunately, sin exists in all realms of human existence, and therefore also in the relationship between human beings and their environment. John Paul II, in the document *Peace with God the Creator*, clearly affirms that the environmental problem is a moral problem.[4] Thus, environmental sin does exist, but it is not the only type of sin, as some environmentalists might suggest. We often focus on the sins of others without looking to our own faults!

The rebellion of Adam and Eve explains the contemporary disrespect for the Creator, which spills over into mistreatment of fellow human beings and recklessness with the rest of creation. As our first parents condemned themselves to death, so do we heedlessly cause truly hazardous ecological situations, leading to the full range of present–day crises. The gravity of sin consists precisely in its undermining and destroying the foundational relationship of God, man and created things. St. Ignatius of Loyola would have each of us sense, taste and feel sin in its horror and destructiveness, and as each one of us is involved, we meditate in the first person. For with my sin I partake of, become one with, a history of de–creation, a story of death and hell.[5]

At the origins of the ecological crisis is the denial of the relationship with God. To be cut off from God is to be separated from the source of life, from the fundamental love and respect for life. When we are so cut off, then we permit ourselves to destroy life and, ecologically speaking, the conditions for life. The environment, instead of being treated with proper respect, becomes subject to irresponsible and violent repression. As the book of Proverbs teaches: 'The virtuous man looks after the lives of his beasts, but the wicked man's heart is ruthless.' (Pr 12:10). By attributing this caring concern to the virtuous, Scripture affirms that a correct relationship with God is reflected in a respectful relationship with the environment, and living rightly within the environment means adoring God in an ordered cosmic hierarchy.[6]

Considering several specific cases, we must keep in mind that a human action is made up of the constitutive elements shown in the figure below. '*Imputability* and responsibility for an action can be diminished or even nullified by ignorance, inadvertence, duress, fear, habit, inordinate attachments, and other psychological or social factors.'[7]

Figure 1

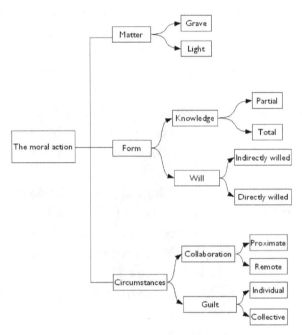

First, following Figure 1, the material element is the act considered in itself, the chosen object. Objectively, the matter can be grave or light. Second, the formal element consists of the awareness of the intellect and the decision of the will, in relation to the end or intention. Awareness can be partial or total. The decision of the will concerns, for example, whether the act is directly or indirectly procured. The third element is circumstantial, or the circumstances of the act. 'The *circumstances*, including the consequences, are secondary elements of the moral act. They contribute to increasing or diminishing the moral goodness or evil of human acts (for example, the amount of a theft). They can also diminish or increase the agent's responsibility (such as acting out of a fear of death). Circumstances of themselves cannot change the moral quality of acts themselves; they can make neither good nor right an action that is in itself evil.'[8] For example, collaboration can be proximate or remote, and guilt can be collective or individual. It is clear, however, that collective guilt cannot be immediately determined without a clear justification.

In summary, 'A *morally good* act requires the goodness of the object, of the end, and of the circumstances together. An evil end corrupts the action, even if the object is good in itself (such as praying or fasting 'in order to be seen by men'). The *object of choice* can by itself vitiate an act in its entirety. There are some concrete acts ... that it is always wrong to choose, because choosing them entails a disorder of the will, that is, a moral evil.'[9]

These elements can alter the subjective imputability of the sin, or rather whether it is mortal or venial. Mortal sin destroys charity in man's heart due to a grave violation of God's law; it removes man from God, Who is his final end and beatitude, preferring a lesser good to Him. Venial sin allows charity to subsist, though it offends and harms it. In order for a sin to be mortal, three conditions must be satisfied: the object must be grave matter and the sin is also committed with full knowledge and deliberate consent.[10]

Mortal sin presupposes knowledge of the sinful character of the act, of its opposition to God's law. It also implies a consent sufficiently deliberate to be a personal choice. Feigned ignorance and hardness of heart (cf. Mk 3:5–6; Lk 16:19–31) do not diminish, but rather increase, the voluntary character of a sin. *Unintentional ignorance* can diminish or even remove the imputability of a grave offense. But no one is deemed to be ignorant of the principles of the moral law, which are written in the conscience of every man. The promptings of feelings and passions can also diminish the voluntary and free character of the offence, as can external pressures or pathological disorders. Sin committed through malice, by deliberate choice of evil, is the gravest.[11]

It must also be considered that sometimes the collateral effects of a given product or action are unknown in a given historical moment, and are ascertained only after more extensive research. It may be that at one point in time people acted in good faith, and only later were problems averred. Some examples of this growing awareness are the harmful effects of mercury, radioactivity, DDT, asbestos and microwaves.

The relation between morality and civil law must also be taken into consideration. Here, as John Paul II insisted, the need for civil law to conform to the moral law must be affirmed.[12] This teaching is continuous with the entire tradition of the Church, and was also taught by Pope John XXIII:

> Governmental authority, therefore, is a postulate of the moral order and derives from God. Consequently, laws and decrees passed in contravention of the moral order, and hence of the divine will, can have no binding force in conscience ... Indeed, the passing of such laws undermines the very nature of authority and results in shameful abuse.[13]

This is also the clear teaching of St Thomas Aquinas, who wrote:

> ...human law has the rationale of law in so far as it is in accordance with right reason, and as such it obviously derives from eternal law. A law which is at variance with reason is to that extent unjust and

has no longer the rationale of law. It is rather an act
of violence.[14]

The Angelic Doctor insists that every human law enjoys
the true nature of law, insofar as it is derived from the
natural law. However, if in any point it conflicts with the
natural law, it is no longer a law but a perversion of law.[15]

The first case to consider, then, is that of a worker in a
factory or company that creates pollution. In reference to
the matter, pollution can be more or less grave depending
on its quantity and quality. Regarding the quantity, it is
apparent that the more concentrated the discharged
substances are, the more serious the damage is. For example, in the case of soap discharged into a river, if it is
diluted, the damage is rapidly repaired. Balance is rapidly
reestablished. The proportion of concentration to repara-
bility, however, is not often so simple. Over a certain
threshold of pollution, the damage might be irreparable.
As far as the quality is concerned, some substances are
more poisonous than others: for example, radioactive
waste is more dangerous than soap, also because it has
long–lasting effects. Likewise with non–poisonous
substances: some are less biodegradable, with considerable
differences between paper, plastic and glass.

Clearly it is much graver to dump substances which
cause damage over large time periods than ones which
have only a brief impact. It is also grave to dump
substances which last a long time in a chain process: for
example, mercury, when it ends up in water, is absorbed
by and accumulates in fish, which are then eaten by
people. It must be determined whether the damage
concerns human beings directly or indirectly, through the
food chain.

Concerning the form, awareness must be taken into consideration. It is sometimes true that the effects of a substance dumped into a river might not yet be known. For example, the negative effects of radioactivity were not always known in the past. We only now know that ionizing radiation can damage living cells. The most serious cellular damage leads to the modification of inheritable material. Changes to the DNA and in the chromosomal structure can be transmitted to future cell generations.

In the interest of radiation protection, it is normal to recall that any dose — however small — of radiation has a small possibility of causing damage. Given the experience of recent years, it is necessary that great care be taken with the waste. Another element to consider is that a given employee may be less aware of such effects than the department head or manager. Those who have greater awareness can better avoid the damage, and consequently they are guiltier in the case of transgressions. Lack of attention and laziness also perhaps play a role, as well as greediness for illicit profit.

Concerning the will, sometimes an act of pollution is not directly willed but is the effect of negligence, which is also a form of guilt. Sometimes a polluting act results from a wrong decision, such as when a waste purification system is not implemented because the cost would lead to the company shutting down and laying off its workers. Here it should be recalled that a good intention (for example, helping one's neighbour) does not make an action which is in itself wrong either good or evil. The end does not justify the means.[16] Sometimes, however, greed is a temptation which leads to the implementation of inadequate purification systems.

In the case of Chernobyl, the disaster was caused by an unauthorized procedure (which was also carried out incorrectly). It was therefore much graver, because it almost directly caused the disaster. In general, in order to evaluate the circumstances, the historical facts must be taken into consideration.

Sometimes an environmental problem is the result of various sources of pollution (for example in seas and rivers); other times it is only one company that is guilty. Guilt can therefore be collective or individual. If it is collective, usually some individuals are guiltier than others. However, if the guilt is collective, the gravity of the action is not automatically reduced for all the individuals or legal persons involved. Here it is useful to consider the analogy of a robbery committed by an individual or a group of individuals (in criminal law, a 'conspiracy to commit a crime' constitutes an 'aggravating circumstance' if it does not constitute a related crime).

Cooperation or collaboration in the act of polluting can be considered proximate or remote. Those who help the ones who decide or carry out the act itself are 'proximate agents' in the act. Those who work in the same company, if they are in direct relation to the act, are potentially 'remote agents.' Can it be said, however, that even those who sell the products made by this company remotely cooperate in the act? It is necessary to trace at least an indirect line of causality in order to respond affirmatively. This argument is important for those who protest against companies guilty of polluting.

We are therefore dealing with a question of responsibility. All of these factors and others are important for judging the gravity of the environmental sin and also, at the level of local law, for judging the gravity of a crime or civil responsibility in this area.

Clearly, even an act of negligence — whether qualitative or quantitative — concerning a recognizably grave matter constitutes a mortal sin. The difficulty in the case of confessing it comes in the moment of determining the satisfaction (penance), which includes restitution. Every grave act which does harm to another person or to the common good demands not only a penance, but also a certain form of reparation. Perhaps if a person confesses having caused such an act, in the future they should try to repair the damage in some way. If radioactive waste is concerned, for example, perhaps the person should make reparations toward those who are sick with cancer due to that pollution.

Wasting is also a sin which has an individual and collective dimension. The remedy is a greater awareness of the problem and of the possible solutions, implying a greater awareness of the need to not waste resources and to develop methods for reusing waste. Scientists and engineers believe that there is a real possibility of considerably reducing the current pollution level and slowly eliminating the mountains of refuse. Drawing raw materials out of refuse and harmful materials has a dual aim: on the one hand, the reduction of the quantity of material used, and on the other hand, the elimination of often toxic chemical waste. An intelligent way of reducing waste could be the implementation of paper and glass recycling, the use of biodegradable plastics and the recovery of materials from unusable computer, car and airplane parts.

Another form of interference with the social environment, though atypical, comes from private use of computers in human society and associated crimes. Today,

since computers have become part of our artificial environment, not only fraud but also hacking constitute real and
serious infringements on human life. Since today it is
possible to commit certain sins by simply pressing a button
on a computer, there is a tendency to think that the matter
is no longer grave. However, it must be taken into consideration that this is not just a more or less entertaining
informational game. It is clear to everyone, for example,
that fraud or robbery through the computer is objectively
grave in function of the sum of money that was unjustly
appropriated.

Computer abuse is a topic which has inspired many
movies. There are many situations that affect our human
environment and culture. The confidentiality of medical
documents is just one example. Entering through a
computer into a system where personal data are stored
with the aim of acquiring private information for profit,
curiosity or fun, is objectively illicit, since everything
bound to the 'privacy' of the person, though it may be
immaterial, is a good which must be respected. There have
been some cases in which students entered into their
university's computer system in order to change their
grades!

In some situations, the authorities charged with public
safety must necessarily intervene, with appropriate guarantees (authorization by the courts), in the 'private' sector
of determined individuals in order to prevent or suppress
crimes.

Then there are acts of vandalism carried out through
so–called viruses. These are programs which, entering into
other systems, compromise operative efficiency. There
have been cases where many months of work were

destroyed by such actions; this is also a grave matter on the moral level.

Clearly, in the moral realm, full account must be taken of the work of the Holy Spirit Who guides the minds and hearts of men and women to make the right decisions concerning God's gifts:

The outpouring of the Holy Spirit enlightens the thoughts of artists, poets, and scientists. Their great minds receive from You prophetic insights into Your laws, and reveal to us the depth of Your creative wisdom. Unwittingly, their works speak of You; how great You are in all You have created, how great You are in man!

Glory to You, showing your unfathomable might in the laws of the universe!

Glory to You, for all nature is permeated by Your laws,

Glory to You for what you have revealed to us in Your goodness,

Glory to You for all that remains hidden from us in Your wisdom,

Glory to You for the inventiveness of the human mind,

Glory to You for the invigorating effort of work,

Glory to You for the tongues of fire which bring inspiration,

Glory to You, O God, from age to age.[17]

5.2 Ideas for an environmental spirituality

The starting point for spirituality is sound doctrine, drawn from Scripture and Tradition. The consequence of such doctrine is clear: man's relation with God determines man's relation with others and with his environment. This is why Christian culture has always recognized the creatures surrounding man as God's gifts, to be cultivated and cared for with a sense of gratitude toward the Creator. In particular, Benedictine and Franciscan spirituality have witnessed to this sort of relationship between man and the created environment, nurturing in him an attitude of respect toward all the realities of the surrounding world.[18]

Today, we often witness an array of exasperated opposing positions, which are not useful for Christian spirituality. On the one hand, widespread birth control policies are demanded in the name of the exhaustibility and insufficiency of environmental resources, particularly for poor or developing countries. On the other hand, in the name of an idea aspiring to ecocentrism and biocentrism, it is proposed to eliminate the ontological and axiological difference between humans and other living beings, considering the biosphere a biotic unit of homogeneous value. Thus, man's greater responsibility is eliminated in favour of an egalitarian consideration of the 'dignity' of all living beings.[19] These false attempts to reduce human dignity are fostered by the errors of so–called creation spirituality, which as has been seen earlier tends to pantheism and to religious indifferentism where all religions are seen as having equal saving power. In this context, there is a lack of a true distinction between matter and spirit as well as a neglect for the specificity of the Christian doctrine on creation. These errors devalue the doctrines of the Incarnation and Redemption with respect to the doctrine of creation.[20]

5.2.1 A biblical approach

This fine approach is expressed in Robert Faricy's reflections on an environmental spirituality.[21] The starting point is in the Song of Songs, where the following is read:

> I came down to the nut garden to look at the fresh growth of the valley, to see if the vines were in bloom, if the pomegranates had blossomed. (Sg 6:11)

I too can 'come down to the nut garden to look at the vineyards and flowers'; I can find myself, with my imagination, alongside the prince and the bride. Our interpretation of this passage — for an environmental spirituality — is tropological; it is different from the literal interpretation of many exegetes and from the allegorical interpretation whereby the bride is the Church and the prince is Christ (which Origen, Gregory of Nyssa and Saint Bernard of Clairvaux sustained).[22] Together with Saint John of the Cross, we see the Song as a metaphor for our relationship with the world. The garden is nature, the prince is Christ and each of us is the bride.

One can ponder six fundamental ways of finding God in nature: in service, praise, thanksgiving, contemplation, meditation and considering nature a metaphor of Jesus Christ in His humanity. In this regard, the Western Christians conceive of the cosmos as a house, whereas the Eastern Christians understand the cosmos as a temple.

First, regarding service, the following may be affirmed:

> For man, created to God's image, received a mandate to subject to himself the earth and all it contains, and to govern the world with justice and holiness; a mandate to relate himself and the totality of things to Him Who was to be acknowledged as the Lord and Creator of all. Thus, by the subjection of all things to man, the name of God would be wonderful in all the earth. This mandate concerns the whole of everyday activity as well. For while providing the substance of life for themselves and their families, men and women ... can justly consider that by their labour they are unfolding the Creator's work.[23]

In this way, human collaboration with divine Providence can be seen. Exploiting nature as if it were a merely a mine, rather than caring for it and keeping in mind that it is a garden, we have used it inappropriately. We extend the work of creation when we carry out service which redeems nature, service which is part of the logic of the Cross. We recall that Jesus suffered, sweated blood, and underwent agony in the Garden of Olives in order to redeem the sin committed in the Garden of Eden.

Second, we can praise the Lord directly for the wonders and beauties of nature which He has given us, offering them directly to Him for love of Him. We find some examples of this type of praise in the psalms of the Old Testament, for example in Psalm 148:

> Praise him sun and moon;
> give praise, all shining stars.
> Praise him, highest heavens,
> you waters above the heavens...
> Praise the Lord from the earth,
> you sea monsters and all deep waters;
> You lightning and hail, snow and clouds,
> storm winds that fulfill his command;
> You mountains and all hills,
> fruit trees and all cedars;
> You animals wild and tame,
> you creatures that crawl and fly...
> Let them all praise the Lord's name.

The First Vatican Council declared that 'the world was created for the glory of God.'[24] The glory of God includes the glory that God gives to His creatures and for which they manifest His greatness, as well as the praise (glory) that we should render to God for the manifestations of His greatness, such as in nature. Praise is not the same as thanksgiving. When we thank God, we show our gratitude for His gifts which by our thanksgiving we attribute to Him; but when we praise God, we acclaim Him not for His gifts, but for Himself. Praise is therefore the point in which thanksgiving becomes giving thanks to God for being God. Praise is similar to adoration, but praise is more active and extroverted. Praise limits itself to acclaiming and applauding God in His creation as well, because He is the Lord of His creation. It does not look to the past, like thanksgiving, nor to the future, like prayers of petition, but rather looks directly at the Lord and applauds. As human beings, we represent all of creation, and thereby present God with praise on behalf of all of creation. We gather all beings together (animals, plants) in praise of God, like in the *Canticle of the Creatures* of St. Francis:

> We praise You, Lord, for all Your creatures,
> especially for Brother Sun,
> who is the day through whom You give us light.
> And he is beautiful and radiant with great splendour,
> of You Most High, he bears your likeness.
> We praise You, Lord, for Sister Moon and the stars,
> in the heavens you have made them bright,
> precious and fair.

The third way is thanksgiving, which has the Most Holy Eucharist at its apex. Service and praise lead us to thank God for His gifts in nature, and converge above all on His gift of Himself in the Most Holy Eucharist. In fact, service, praise and thanksgiving all combine when we partake of the Lord's Supper, the Eucharist. We bless God for all of creation, because through His goodness we have bread to offer — fruit of the earth and work of human hands — and we have wine to offer — fruit of the vine and work of human hands. At the moment of consecration, the Eucharist celebrates the divinizing transformation — carried out by the Father through the Son in the Holy Spirit — of our work upon nature as the work of human hands becomes the Body and Blood of Christ. The Eucharistic model points toward God, and not toward nature: it is therefore a strong remedy against cosmocentrism.

Praise is found in the entire Eucharistic celebration. In the preface of the Fourth Eucharistic Prayer, we praise the Creator 'in the name of every creature under heaven.' This Eucharistic Prayer praises God because He has formed us in His likeness, He has placed us above all of nature to serve Him, our Creator, and to govern all creatures. The entire Eucharistic celebration is a prayer of thanksgiving. All of nature, represented by the bread and wine (nature transformed by man), is offered to the Father, then transformed by the power of His Son Jesus, in order to be offered anew. The Eucharist provides us with the exemplary model of how to thank the Father for nature. In devout recognition, we offer it to Him through the Son, in the Holy Spirit: 'Through Him, with Him, in Him, in the unity of the Holy Spirit, all glory and honour is Yours Almighty Father, for ever and ever. Amen.'

The fourth way is contemplation of God through nature. For Western Christians, this contemplation begins with creation and moves toward God; for Eastern Christians, contemplation begins with God and moves toward creation. Consider the example of Psalm 104:

Bless the Lord, my soul! ...
How varied are your works, Lord!
In wisdom you have wrought them all;
the earth is full of your creatures.
Look at the sea, great and wide! ...
Bless the Lord, my soul! Hallelujah!

Psalm 104 moves from an initial burst of praise to the contemplation of God through and in nature, and ends with another burst of praise. Job's contemplation (chapters 38–41), on the other hand, begins in destitution and depression, passing to the contemplation of God in the wonders of His creation and seeing in it the transcendent wisdom of the Creator which goes well beyond what we can comprehend. Contemplation of the Lord in and through nature concludes in humility, as expressed by Job in his prayer:

> I had heard you by word of mouth, but now my eye has seen you. Therefore I disown what I have said, and repent in dust and ashes. (Jb 42:5–6)

In the Letter to the Romans, Paul criticizes the pagans who have not turned to God despite having seen Him in nature, which surrounds them (cf. Rm 1:18, 32). Despite having seen Him in nature, they have not turned to Him in love. Contemplating God through nature is therefore this: turning to Him and gazing with love upon His attributes, manifested in nature.

In Eastern Christendom, the Holy Spirit is praised for the beauty of His creation:

> In the strength of the Holy Spirit each flower gives out its scent – sweet perfume, delicate colour, beauty of the whole universe revealed in the tiniest thing. Glory and honour to God the Giver of life, who covers the fields with their carpet of flowers, crowns the plains with harvest of gold and the blue of corn–flowers, and our souls with the joy of contemplating him. O be joyful and sing to him: Alleluia![25]

In the fifth way, we see how meditation on nature leads to God. Meditation represents our effort, while contemplation is a gift from God. We can reflect on nature and thereby reach some conclusions about God. Reflecting on what we see in nature, we can learn more about nature's Author. We will now consider two examples.

First, the presence of the mystery of nature leads to a reflection which ends with a greater comprehension of the mysteriousness and incomprehensibility of God and His ways:

> As the heavens tower over the earth,
> so God's love towers over the faithful.
> As far as the east is from the west,
> so far have our sins been removed from us. (Ps 103:11–12)

Second, we can deepen our knowledge of the faithfulness of the Lord's love by meditating on nature, which He created:

> Though the mountains leave their place
> and the hills be shaken,
> My love shall never leave you... (Is 54:10)

The sixth way regards nature as a metaphor for the humanity of Christ. In His teaching, Christ makes use of natural images, sometimes as a metaphor for Himself, such as when He says 'I am the light of the world' (Jn 8:12), or 'I am the true vine' (Jn 15:1). We pray to Christ in the Sacred Heart, identified with His human heart, symbol and natural seat of His love.

Some poets see metaphors for Jesus Christ in nature, such as Joseph Mary Plunkett (1887–1916) in his verse 'I see His Blood upon the Rose':

> I see his blood upon the rose

And in the stars the glory of his eyes,
His body gleams amid eternal snows,
His tears fall from the skies.

I see his face in every flower;
The thunder and the singing of the birds
Are but his voice—and carven by his power
Rocks are his written words.

All pathways by his feet are worn,
His strong heart stirs the ever–beating sea,
His crown of thorns is twined with every thorn,
His cross is every tree.

Similarly, in the Eastern tradition, images are adopted from nature to describe the sweetness of the Word:

How filled with sweetness are those whose thoughts dwell on you: how life–giving your holy Word; to speak with you is more soothing than anointing with oil, sweeter than the honeycomb. Praying to you refreshes us and gives us wings: our hearts overflow with warmth; a majesty filled with wisdom permeates nature and all of life! Where you are not, there is only emptiness. Where you are, the soul is filled with abundance, and its song resounds like a torrent of life: Alleluia! [26]

Another example is the hymn to Mary, Mother of God, written by Blessed Hildegard von Bingen (1098–1179), which employs many images from nature:

Hail, o Greenest Branch,
who came forth in the blustering wind of the saints' search.
Hail to you, when the time came and your branches blossomed,
because the warmth of the sun produced in you
a fragrance like balsam.
For a beautiful flower bloomed in you,
which perfumed all the herbs that were dry.

And they all appeared, full and green.
Hence the heavens sent dew to fall on the grass
and the whole earth rejoiced because its womb
brought forth grain
and because the birds of heaven made their nests in it.
Thus food was made for mankind,
and great was the joy of those who ate.
Therefore, Virgin, no joy is lacking in you.
Eve had contempt for all these things.
But now praise be to the Highest.[27]

The Mother of God is also invoked as the *Mystical Rose* and the *Star of the Sea* in her Litany.

5.2.2 The Christian and the world

Another complementary approach could be based on the vocation of the laity understood in terms of the three offices (or *munera*) of the people of God: priestly, prophetic and kingly. This line of thinking includes many of the same elements of the one we just considered:

> The term laity is here understood to mean all the faithful ... [who] are by baptism made one body with Christ and are constituted among the People of God; they are in their own way made sharers in the priestly, prophetical, and kingly functions of Christ; and they carry out for their own part the mission of the whole Christian people in the Church and in the world... But the laity, by their very vocation, seek the kingdom of God by engaging in temporal affairs and by ordering them according to the plan of God. They live in the world, that is, in each and in all of the secular professions and occupations. They live in the ordinary circumstances of family and social life, from which the very web of their existence is woven. They are called there by God that by exercising their proper function and led

by the spirit of the Gospel they may work for the sanctification of the world from within as a leaven. In this way they may make Christ known to others, especially by the testimony of a life resplendent in faith, hope and charity. Therefore, since they are tightly bound up in all types of temporal affairs it is their special task to order and to throw light upon these affairs in such a way that they may come into being and then continually increase according to Christ to the praise of the Creator and the Redeemer.[28]

Regarding the priestly aspect, the faithful offer praise and thanksgiving for creation to the Father (Who created everything) through the Son (through Whom everything was created) in the Holy Spirit (in Whom everything was created):

Low before Him with our praises we fall,
Of Whom, and in Whom, and through Whom are all;
Of Whom, the Father; and in Whom, the Son,
Through Whom, the Spirit, with Them ever One.[29]

Through their priestly office, Christians sanctify the world. The faithful also exercise a prophetic office, teaching by their example how to respect creation and thank God for the gift of the cosmos. In their kingly aspect, instead, Christians are God's vicars (or stewards) within creation who serve and bring Christ's Kingdom to fulfilment .

In the exercise of these three offices, the connection with the four cardinal virtues must also be considered: justice, prudence, temperance and fortitude. One must move from an environmental spirituality to certain more profound attitudes, toward an *ethos* expressed in operative moral habits, such as reverential respect, awe, aesthetic admiration and religious contemplation of nature, gratitude for the gift of life, responsibility and care for every created being. It is very important to provide these attitudes with roots in the classical scheme of the cardinal virtues in order to emphasize four fundamental environmental virtues.

The first step is therefore the subordination of reverential respect to justice, understood in its broad biblical meaning, which requires man to know his proper place in the great order of creation, since God Himself, through His works, has made the world a place for the fullness of life. In this way, the first approach to the world and to the environment must return to being one of awe and reverential fear, as expressed in Psalm 8: 'O Lord, our Lord, how awesome is your name through all the earth!'

The second virtue, prudence, also has environmental relevance. In fact, it encourages the development of a specific knowledge which humanity can apply. It is important to work intensely so as to best understand the state of things, in order to then apply concrete knowledge in the moment of decision–making.

The third virtue, temperance, teaches us to maintain just measure. In this perspective, it is necessary to accept humanity and the world with their respective limits, which are impossible to eliminate. It is crucial, therefore, to preserve the earth and its resources for future generations.

The fourth virtue, fortitude, does not mean just tenacity and resolve in the environmental field, which inspire us to never relent in demanding environmental protection despite setbacks and disappointments, but it also demands the resolve of the human will to survive despite catastrophic predictions of the end of the world, and therefore to actively mobilize all of its energies, including and united with prayer.

In this way, environmental spirituality also involves a journey of conversion. A radical mentality change must include various aspects. First, the recognition of existing connections with the entire ecosystem and living and acting responsibly with awareness of these connections. Next, one undertakes the renunciation of consumption and behaviour habits which harm the natural environment, fostering the establishment of new habits. Also important is the modification of habitual negligence (for example, avoiding superfluous refuse, eliminating or at least reducing mixed waste, and actively recycling: glass, paper, metal and other materials). Another concrete measure is the dutiful attention to measure, moderation, lifestyle discipline, closeness to nature, and care of others, including solidarity with poor people. New capacities should be stimulated to direct creativity towards the environment (involving for example, the virtue of doing things oneself and improvising with limited, simpler, but more environmentally friendly means). Assuming responsibility for creation also means imposing checks and limits, and begins in the family, the household, the neighbourhood and the area where one lives.[30]

The Spiritual Exercises of St Ignatius can facilitate conversion, bring healing to our relationship with the Earth, and enable us to be people of hope, seeking change in cultural attitudes and social structures that contribute to the crisis. In the Third Rule of Discernment, Ignatius affirms on the one hand that we cannot have knowledge of God apart from the created world. At the same time, the created world must be clearly referred to its Creator, thus avoiding any form of cosmocentrism. Ignatius says that consolation is 'an interior movement...aroused in the soul, by which it is inflamed with love of its Creator and Lord, and as a consequence, can love no created thing on the face of the earth for its own sake, but only the Creator of them all.'[31] Despite our abuse of creation, created things continue all along to sing of the mercy of the Lord. St Ignatius invites us to marvel at the heavens, with the sun and moon and all the stars, and the earth with fruit and fish and animals, and to consider how these created things sustain, nourish and protect me, keep me alive and permit me to live and never cease to do so; even when I ignore God and refuse to praise the Divine Mystery, even when I close myself up in isolation from other creatures, even when I refuse to serve Him and abuse these created beings. This reflection on God's mercy invites me to a colloquy with the Merciful One.[32] The First Week of these Exercises enables retreatants to get in touch with the extent of the ecological crisis, but does so in the context of a loving God. We seek healing of our dysfunctional relationships with the Earth, with the humans of the Earth, and with God. In the Second Week, we seek to nourish ourselves in the mystery of God, in the beauty of God, in the presence of Christ upon this the Earth and our soul is nourished by him. The Third Week allows us to confront the suffering of the Earth, the reality of death, and find God there — the suffering Christ. In the Fourth Week, we experience again that the suffering and death are not the end, but life. The life of Christ experienced here on Earth brings us joy, and hope. This is the real gift of the Exercises to the ecological situation. The dynamic of the Exercises brings us to a disposition of hope after prayerfully considering the crisis. Hope–filled compassionate action for the Earth, not paralysis, is the result.

The monasticism of East and West can teach us a great deal through its emphasis on the value of praise, humility, responsible administration ('stewardship'), manual labour, and community. In this perspective, there is no place for egoism or cosmocentrism. The three religious vows of poverty, chastity and obedience are a remedy against the triple concupiscence: 'Christians will have to act by giving a cosmic dimension to their prayer, their listening to the Word, their sacramental life and their asceticism. They will do it by broadly showing the cultural, social and environmental richness that traditional ascetic values possess when opened to history: I am thinking above all ... of voluntary a limitation of needs and a profound liking for all forms of life.'[33] Religious life, furthermore, indicates that this earthly life of ours is not an end in itself, but leads to our future life in a new heaven and a new earth:

When over the earth the light of the setting sun fades away, when the peace of eternal sleep and the quiet of the declining day reign over all, I see Your dwelling–place like tents filled with light, reflected in the shapes of the clouds at dusk: fiery and purple, gold and blue, they speak prophet–like of the ineffable beauty of Your heavenly court, and solemnly call: let us go to the Father!

Glory to You in the quiet hour of evening,

Glory to You, covering the world with deep peace,

Glory to You for the last ray of the setting sun,

Glory to You for the rest of blissful sleep,

Glory to You for Your mercy in the midst of darkness, when the whole world has parted company with us,

Glory to You for the tender emotion of a soul moved to prayer,

Glory to You for the pledge of our awakening on the day which has no evening,

Glory to You, O God, from age to age.[34]

Notes

1   St Augustine, *Contra Faustum manichaeum*, 22 in *PL* 42, 418; St Thomas Aquinas, *Summa Theologiae*, I–II, q.71, a.6. Cf. *CCC* 1849.

2   Cf. *CCC* 1848.

3   Cf. Pope John Paul II, *Address to the Congress on 'Environment and Health'* (24 March 1997), 6.

4   Cf. Pope John Paul II, *Peace with God the Creator, Peace with all of Creation*, 6–7 and 15.

5   See M. Czerny (ed.), 'We live in a broken world' in *Promotio Iustitiae* 70 (1999), 2.2.

6   See *ibid.*.

7   *CCC* 1735.

8   *CCC* 1754.

9   *CCC* 1755.

10   Cf. Pope John Paul II, Apostolic exhortation *Reconciliatio et paenitentia*, 17.

11   *CCC* 1859–1860.

12   Cf. Pope John Paul II, *Evangelium vitae* (25 March 1995), 72. Cf. Idem, *Veritatis Splendor* (6 August 1993), 43.

13   Pope John XXIII, *Pacem in terris* (11 April 1963), 51.

14   St Thomas Aquinas, *Summa Theologiae*, I–II, q. 93, a. 3.

15   See St Thomas Aquinas, *Summa Theologiae*, I–II, q. 95, a. 2. St Thomas cites St Augustine: '*Non videtur esse lex, quae insta non fuerit.*' From *De libero arbitrio*, I, 5, 11 in *PL* 32, 1227.

16   Cf. *CCC* 1753.

17   Metropolitan Tryphon, *An Akathist in Praise of God's Creation*, Ikos 7.

18   Pope John Paul II, *Address to the Congress on 'Environment and Health'* (24 March 1997), 4.

19   *Ibid.*, 5.

20 See chapter 2, pp. 81–88 above where we discussed creation spirituality and its pitfalls.

21 Cf. R. Faricy, *Wind and Sea Obey Him* (London: SCM Press, 1982), Chapter 5.

22 The *tropological* meaning links the person, events or things contained in Scripture with the moral life. While the Beatitudes or the Decalogue speak at the literal, plain level of moral duties, according to the tropological sense, moral lessons can also be learned from things in themselves. Thus, the beauty of Rachel is said to be a sign of the contemplative life, while the ugliness of Leah is a sign of a life marred by vice. The physical appearance of these matriarchs, without reference to their actions, is itself understood as a sign of the moral life.

23 Vatican II, *Gaudium et spes*, 34.

24 Vatican I, Dogmatic Constitution *Dei Filius*, chapter I, canon 5.

25 Metropolitan Tryphon, *An Akathist in Praise of God's Creation*, Kontakion 3.

26 Metropolitan Tryphon, *An Akathist in Praise of God's Creation*, Kontakion 4.

27 Hildegard von Bingen, *De sancta Maria*. The Latin original runs as follows:
O viridissima virga, ave,
quae in ventoso flabro sciscitationis prodisti.
Cum venit tempus, quod tu floruisti in ramis tuis,
ave, ave sit tibi,
quia calor solis in te sudavit sicut odor balsami.
Nam in te floruit pulcher flos,
qui odorem dedit omnibus aromatibus, quae arida erant.
Et illa apparuerunt omnia in viriditate plena.
Unde caeli dederunt rorem super gramen,
et omnis terra laeta facta est,
quonima viscera ipsius frumentum protulerunt,
et quonima volucres caeli nidos in ipsa habuerunt.

Deinde facta est esca hominibus
et gaudium magnum epulantium.
Unde, o suavis Virgo,
in te non deficit ullum gaudium.
Haec omnia Eva contempsit.
Nunc autem laus sit altissimi.

[28] Vatican II, *Lumen gentium*, 31.

[29] From Peter Abelard, 'O quanta qualia':
'Perenni Domino perpes sit gloria,
Ex quo sunt, per quem sunt, in quo sunt omnia;
Ex quo sunt, Pater est, per quem sunt, Filius,
In quo sunt, Patris et Filii Spiritus.'
Translation by J. M. Neale.

[30] Cf. Common Declaration of the Catholic and Evangelical Churches in Germany, *Feeling responsibility for creation* (14 May 1985), 75–77.

[31] St Ignatius of Loyola, *Spiritual Exercises*, 316.

[32] See St Ignatius of Loyola, *Spiritual Exercises*, 61.

[33] Ignatius IV Hazim, Orthodox Patriarch of Antioch, *Trasfigurare la creazione*, (Monastero di Bose: 1994), p. 29.

[34] Metropolitan Tryphon, *An Akathist in Praise of God's Creation*, Ikos 4.

Appendix 1

MESSAGE OF HIS HOLINESS
POPE JOHN PAUL II
FOR THE CELEBRATION OF THE
WORLD DAY OF PEACE
1 JANUARY 1990

PEACE WITH GOD THE CREATOR,
PEACE WITH ALL OF CREATION

Introduction

1. In our day, there is a growing awareness that world peace is threatened not only by the arms race, regional conflicts and continued injustices among peoples and nations, but also by a lack of *due respect for nature*, by the plundering of natural resources and by a progressive decline in the quality of life. The sense of precariousness and insecurity that such a situation engenders is a seedbed for collective selfishness, disregard for others and dishonesty.

Faced with the widespread destruction of the environment, people everywhere are coming to understand that we cannot continue to use the goods of the earth as we have in the past. The public in general as well as political leaders are concerned about this problem, and experts from a wide range of disciplines are studying its causes. Moreover, a new *ecological awareness* is beginning to emerge which, rather than being downplayed, ought to be encouraged to develop into concrete programmes and initiatives.

2. Many ethical values, fundamental to the development of a *peaceful society*, are particularly relevant to the ecological question. The fact that many challenges facing the world today are interdependent confirms the need for carefully coordinated solutions based on a morally coherent world view.

For Christians, such a world view is grounded in religious convictions drawn from Revelation. That is why I should like to begin this Message with a reflection on the biblical account of creation. I would hope that even those who do not share these same beliefs will find in these pages a common ground for reflection and action.

I. 'And God saw that it was good'

3. In the Book of Genesis, where we find God's first self–revelation to humanity (*Gen* 1–3), there is a recurring refrain: '*And God saw that it was good*'. After creating the heavens, the sea, the earth and all it contains, God created man and woman. At this point the refrain changes markedly: 'And God saw everything that he had made, and behold, *it was very good* (*Gen* 1:31). God entrusted the whole of creation to the man and woman, and only then – as we read – could he rest 'from all his work' (*Gen* 2:3).

Adam and Eve's call to share in the unfolding of God's plan of creation brought into play those abilities and gifts which distinguish the human being from all other creatures. At the same time, their call established a fixed relationship between mankind and the rest of creation. Made in the image and likeness of God, Adam and Eve were to have exercised their dominion over the earth (*Gen* 1:28) with wisdom and love. Instead, they destroyed the existing harmony *by deliberately going against the Creator's plan*, that is, by choosing to sin. This resulted not only in man's alienation from himself, in death and fratricide, but also in the earth's 'rebellion' against him (cf. *Gen* 3:17–19; 4:12). All of creation became subject to futility, waiting in a mysterious way to be set free and to obtain a glorious liberty together with all the children of God (cf. *Rom* 8:20–21).

4. Christians believe that the Death and Resurrection of Christ accomplished the work of reconciling humanity to the Father, who 'was pleased ... through (Christ) to reconcile to himself *all things*, whether on earth or in heaven, making peace by the blood of his cross' (*Col* 1:19–20). Creation was thus made new (cf. *Rev* 21:5). Once subjected to the bondage of sin and decay (cf. *Rom* 8:21), it has now received new life while 'we wait for new heavens and a new earth in which righteousness dwells' (2 *Pt* 3:13). Thus, the Father 'has made known to us in all wisdom and insight the mystery...which he set forth in Christ as a plan for the fulness of time, to unite *all things* in him, all things in heaven and things on earth' (*Eph* 1:9–10).

5. These biblical considerations help us to understand better *the relationship between human activity and the whole of creation*. When man turns his back on the Creator's plan, he provokes a disorder which has inevitable repercussions on the rest of the created order. If man is not at peace with God, then earth itself cannot be at peace: 'Therefore the land mourns and all who dwell in it languish, and also the

beasts of the field and the birds of the air and even the fish of the sea are taken away' (*Hos* 4:3).

The profound sense that the earth is 'suffering' is also shared by those who do not profess our faith in God. Indeed, the increasing devastation of the world of nature is apparent to all. It results from the behaviour of people who show a callous disregard for the hidden, yet perceivable requirements of the order and harmony which govern nature itself .

People are asking anxiously if it is still possible to remedy the damage which has been done. Clearly, an adequate solution cannot be found merely in a better management or a more rational use of the earth's resources, as important as these may be. Rather, we must go to the source of the problem and face in its entirety that profound moral crisis *of which the destruction of the environment is only one troubling aspect.*

II. The ecological crisis: a moral problem

6. Certain elements of today's ecological crisis reveal its moral character. First among these is the *indiscriminate* application of advances in science and technology. Many recent discoveries have brought undeniable benefits to humanity. Indeed, they demonstrate the nobility of the human vocation to participate *responsibly* in God's creative action in the world. Unfortunately, it is now clear that the application of these discoveries in the fields of industry and agriculture have produced harmful long–term effects. This has led to the painful realization that *we cannot interfere in one area of the ecosystem without paying due attention both to the consequences of such interference in other areas and to the well–being of future generations.*

The gradual depletion of the ozone layer and the related 'greenhouse effect' has now reached crisis proportions as a consequence of industrial growth, massive urban concentrations and vastly increased energy needs. Industrial waste, the burning of fossil fuels, unrestricted deforestation, the use of certain types of herbicides, coolants and propellants: all of these are known to harm the atmosphere and environment. The resulting meteorological and atmospheric changes range from damage to health to the possible future submersion of low–lying lands.

While in some cases the damage already done may well be irreversible, in many other cases it can still be halted. It is necessary, however, that the entire human community – individuals, States and international bodies – take seriously the responsibility that is theirs.

7. The most profound and serious indication of the moral implications underlying the ecological problem is the lack of *respect for life* evident in many of the patterns of environmental pollution. Often, the interests of production prevail over concern for the dignity of workers, while economic interests take priority over the good of individuals and even entire peoples. In these cases, pollution or environmental destruction is the result of an unnatural and reductionist vision which at times leads to a genuine contempt for man.

On another level, delicate ecological balances are upset by the uncontrolled destruction of animal and plant life or by a reckless exploitation of natural resources. It should be pointed out that all of this, even if carried out in the name of progress and well–being, is ultimately to mankind's disadvantage.

Finally, we can only look with deep concern at the enormous possibilities of biological research. We are not yet in a position to assess the biological disturbance that could result from indiscriminate genetic manipulation and from the unscrupulous development of new forms of plant and animal life, to say nothing of unacceptable experimentation regarding the origins of human life itself. It is evident to all that in any area as delicate as this, indifference to fundamental ethical norms, or their rejection, would lead mankind to the very threshold of self–destruction.

Respect for life, and above all for the dignity of the human person, is the ultimate guiding norm for any sound economic, industrial or scientific progress.

The complexity of the ecological question is evident to all. There are, however, certain underlying principles, which, while respecting the legitimate autonomy and the specific competence of those involved, can direct research towards adequate and lasting solutions. These principles are essential to the building of a peaceful society; *no peaceful society can afford to neglect either respect for life or the fact that there is an integrity to creation.*

III. In search of a solution

8. Theology, philosophy and science all speak of a harmonious universe, of a 'cosmos' endowed with its own integrity, its own internal, dynamic balance. *This order must be respected.* The human race is called to explore this order, to examine it with due care and to make use of it while safeguarding its integrity.

On the other hand, the earth is ultimately *a common heritage, the fruits of which are for the benefit of all.* In the words of the Second Vatican Council, 'God destined the earth and all it contains for the use of every individual and all peoples'.[1] This has direct consequences for the problem at hand. It is manifestly unjust that a privileged few should continue to accumulate excess goods, squandering available resources, while masses of people are living in conditions of misery at the very lowest level of subsistence. Today, the dramatic threat of ecological breakdown is teaching us the extent to which greed and selfishness – both individual and collective – are contrary to the order of creation, an order which is characterized by mutual interdependence.

9. The concepts of an ordered universe and a common heritage both point to the necessity of a *more internationally coordinated approach to the management of the earth's goods.* In many cases the effects of ecological problems transcend the borders of individual States; hence their solution cannot be found solely on the national level. Recently there have been some promising steps towards such international action, yet the existing mechanisms and bodies are clearly not adequate for the development of a comprehensive plan of action. Political obstacles, forms of exaggerated nationalism and economic interests – to mention only a few factors – impede international cooperation and long–term effective action.

The need for joint action on the international level *does not lessen the responsibility of each individual State.* Not only should each State join with others in implementing internationally accepted standards, but it should also make or facilitate necessary socio–economic adjustments within its own borders, giving special attention to the most vulnerable sectors of society. The State should also actively endeavour within its own territory to prevent destruction of the atmosphere and biosphere, by carefully monitoring, among other things, the impact of new technological or scientific advances. The State also has the responsibility of ensuring that its citizens are not exposed to dangerous pollutants or toxic wastes. *The right to a safe environment* is ever more insistently presented today as a right that must be included in an updated Charter of Human Rights.

IV. The urgent need for a new solidarity

10. The ecological crisis reveals the *urgent moral need for a new solidarity*, especially in relations between the developing nations and those that are highly industrialized. States must increasingly share responsibility, in complimentary ways, for the promotion of a natural and social environment that is both peaceful and healthy. The newly industrialized States cannot, for example, be asked to apply restrictive environmental standards to their emerging industries unless the industrialized States first apply them within their own boundaries. At the same time, countries in the process of industrialization are not morally free to repeat the errors made in the past by others, and recklessly continue to damage the environment through industrial pollutants, radical deforestation or unlimited exploitation of non–renewable resources. In this context, there is urgent need to find a solution to the treatment and disposal of toxic wastes.

No plan or organization, however, will be able to effect the necessary changes unless world leaders are truly convinced of the absolute need for this new solidarity, which is demanded of them by the ecological crisis and which is essential for peace. *This need presents new opportunities for strengthening cooperative and peaceful relations among States.*

11. It must also be said that the proper ecological balance will not be found without *directly addressing the structural forms of poverty* that exist throughout the world. Rural poverty and unjust land distribution in many countries, for example, have led to subsistence farming and to the exhaustion of the soil. Once their land yields no more, many farmers move on to clear new land, thus accelerating uncontrolled deforestation, or they settle in urban centres which lack the infrastructure to receive them. Likewise, some heavily indebted countries are destroying their natural heritage, at the price of irreparable ecological imbalances, in order to develop new products for export. In the face of such situations it would be wrong to assign responsibility to the poor alone for the negative environmental consequences of their actions. Rather, the poor, to whom the earth is entrusted no less than to others, must be enabled to find a way out of their poverty. This will require a courageous reform of structures, as well as new ways of relating among peoples and States.

12. But there is another dangerous menace which threatens us, namely *war*. Unfortunately, modern science already has the capacity to change the environment for hostile purposes. Alterations of this kind over the long term could have unforeseeable and still more serious consequences. Despite the international agreements which prohibit chemical, bacteriological and biological warfare, the fact is that laboratory research continues to develop new offensive weapons capable of altering the balance of nature.

Today, any form of war on a global scale would lead to incalculable ecological damage. But even local or regional wars, however limited, not only destroy human life and social structures, but also damage the land, ruining crops and vegetation as well as poisoning the soil and water. The survivors of war are forced to begin a new life in very difficult environmental conditions, which in turn create situations of extreme social unrest, with further negative consequences for the environment.

13. Modern society will find no solution to the ecological problem unless it *takes a serious look at its life style*. In many parts of the world society is given to instant gratification and consumerism while remaining indifferent to the damage which these cause. As I have already stated, the seriousness of the ecological issue lays bare the depth of man's moral crisis. If an appreciation of the value of the human person and of human life is lacking, we will also lose interest in others and in the earth itself. Simplicity, moderation and discipline, as well as a spirit of sacrifice, must become a part of everyday life, lest all suffer the negative consequences of the careless habits of a few.

An education in ecological responsibility is urgent: responsibility for oneself, for others, and for the earth. This education cannot be rooted in mere sentiment or empty wishes. Its purpose cannot be ideological or political. It must not be based on a rejection of the modern world or a vague desire to return to some 'paradise lost'. Instead, a true education in responsibility entails a genuine conversion in ways of thought and behaviour. Churches and religious bodies, non–governmental and governmental organizations, indeed all members of society, have a precise role to play in such education. The first educator, however, is the family, where the child learns to respect his neighbour and to love nature.

14. *Finally, the aesthetic value of creation cannot be overlooked.* Our very contact with nature has a deep restorative power; contemplation of its magnificence imparts peace and serenity. The Bible speaks again and again of the goodness

and beauty of creation, which is called to glorify God (cf. *Gen* 1:4ff; *Ps* 8:2; 104:1ff; *Wis* 13:3–5; *Sir* 39:16, 33; 43:1, 9). More difficult perhaps, but no less profound, is the contemplation of the works of human ingenuity. Even cities can have a beauty all their own, one that ought to motivate people to care for their surroundings. Good urban planning is an important part of environmental protection, and respect for the natural contours of the land is an indispensable prerequisite for ecologically sound development. The relationship between a good aesthetic education and the maintenance of a healthy environment cannot be overlooked.

V. *The ecological crisis: a common responsibility*

15. Today the ecological crisis has assumed such proportions as to be *the responsibility of everyone*. As I have pointed out, its various aspects demonstrate the need for concerted efforts aimed at establishing the duties and obligations that belong to individuals, peoples, States and the international community. This not only goes hand in hand with efforts to build true peace, but also confirms and reinforces those efforts in a concrete way. When the ecological crisis is set within the broader context of *the search for peace* within society, we can understand better the importance of giving attention to what the earth and its atmosphere are telling us: namely, that there is an order in the universe which must be respected, and that the human person, endowed with the capability of choosing freely, has a grave responsibility to preserve this order for the well–being of future generations. I wish to repeat that *the ecological crisis is a moral issue*.

Even men and women without any particular religious conviction, but with an acute sense of their responsibilities for the common good, recognize their obligation to contribute to the restoration of a healthy environment. All the more should men and women who believe in God the Creator, and who are thus convinced that there is a well–defined unity and order in the world, feel called to address the problem. Christians, in particular, realize that their responsibility within creation and their duty towards nature and the Creator are an essential part of their faith. As a result, they are conscious of a vast field of ecumenical and interreligious cooperation opening up before them.

16. At the conclusion of this Message, I should like to address directly my brothers and sisters in the Catholic Church, in order to remind them of their serious obligation to care for all of creation. The commitment of believers to a healthy environment for everyone stems directly from their belief in God the Creator, from their recognition of the effects of original and personal sin, and from the certainty of having been redeemed by Christ. Respect for life and for the dignity of the human person extends also to the rest of creation, which is called to join man in praising God (cf. *Ps* 148:96).

In 1979, I proclaimed Saint Francis of Assisi as the heavenly Patron of those who promote ecology.[2] He offers Christians an example of genuine and deep respect for the integrity of creation. As a friend of the poor who was loved by God's creatures, Saint Francis invited all of creation – animals, plants, natural forces, even Brother Sun and Sister Moon – to give honour and praise to the Lord. The poor man of Assisi gives us striking witness that when we are at peace with God we are better able to devote ourselves to

building up that peace with all creation which is inseparable from peace among all peoples.

It is my hope that the inspiration of Saint Francis will
help us to keep ever alive a sense of 'fraternity' with all
those good and beautiful things which Almighty God has
created. And may he remind us of our serious obligation to
respect and watch over them with care, in light of that
greater and higher fraternity that exists within the human
family.

Notes

1   Vatican II, *Gaudium et Spes*, 69.
2   Cf. Pope John Paul II, Apostolic Letter *Inter Sanctos* in *AAS*
     71 (1979), 1509f..

Appendix 2

Declaration on the Environment

POPE JOHN PAUL II

AND

PATRIARCH BARTHOLOMEW I OF CONSTANTINOPLE

We are gathered here today in the spirit of peace for the good of all human beings and for the care of creation. At this moment in history, at the beginning of the third millennium, we are saddened to see the daily suffering of a great number of people from violence, starvation, poverty, and disease. We are also concerned about the negative consequences for humanity and for all creation resulting from the degradation of some basic natural resources such as water, air and land, brought about by an economic and technological progress which does not recognize and take into account its limits.

Almighty God envisioned a world of beauty and harmony, and He created it, making every part an expression of His freedom, wisdom and love (cf. Gen 1:1-25).

At the centre of the whole of creation, He placed us, human beings, with our inalienable human dignity. Although we share many features with the rest of the living beings, Almighty God went further with us and gave us an immortal soul, the source of self-awareness and freedom, endowments that make us in His image and likeness (cf. Gen 1:26-31; 2:7). Marked with that resemblance, we have been placed by God in the world in order to cooperate with Him in realizing more and more fully the divine purpose for creation.

At the beginning of history, man and woman sinned by disobeying God and rejecting His design for creation. Among the results of this first sin was the destruction of the original harmony of creation. If we examine carefully the social and environmental crisis which the world community is facing, we must conclude that we are still betraying the mandate God has given us: to be stewards called to collaborate with God in watching over creation in holiness and wisdom.

God has not abandoned the world. It is His will that His design and our hope for it will be realized through our cooperation in restoring its original harmony. In our own time we are witnessing a growth of an ecological awareness which needs to be encouraged, so that it will lead to practical programs and initiatives. An awareness of the relationship between God and humankind brings a fuller sense of the importance of the relationship between human beings and the natural environment, which is God's creation and which God entrusted to us to guard with wisdom and love (cf. Gen 1:28).

Respect for creation stems from respect for human life and dignity. It is on the basis of our recognition that the world is created by God that we can discern an objective moral order within which to articulate a code of environmental ethics. In this perspective, Christians and all other believers have a specific role to play in proclaiming moral values and in educating people in ecological awareness, which is none other than responsibility towards self, towards others, towards creation.

What is required is an act of repentance on our part and a renewed attempt to view ourselves, one another, and the world around us within the perspective of the divine design for creation. The problem is not simply economic and technological; it is moral and spiritual. A solution at the economic and technological level can be found only if we undergo, in the most radical way, an inner change of heart, which can lead to a change in lifestyle and of unsustainable patterns of consumption and production. A genuine conversion in Christ will enable us to change the way we think and act.

First, we must regain humility and recognize the limits of our powers, and most importantly, the limits of our knowledge and judgment. We have been making decisions, taking actions, and assigning values that are leading us away from the world as it should be, away from the design of God for creation, away from all that is essential for a healthy planet and a healthy commonwealth of people. A new approach and a new culture are needed, based on the centrality of the human person within creation and inspired by environmentally ethical behaviour stemming from our triple relationship to God, to self, and to creation. Such an ethics fosters interdependence and stresses the principles of universal solidarity, social justice, and responsibility, in order to promote a true culture of life.

Secondly, we must frankly admit that humankind is entitled to something better than what we see around us. We and, much more, our children and future generations are entitled to a better world, a world free from degradation, violence and bloodshed, a world of generosity and love.

Thirdly, aware of the value of prayer, we must implore God the Creator to enlighten people everywhere regarding the duty to respect and carefully guard creation.

We therefore invite all men and women of good will to ponder the importance of the following ethical goals:

1. To think of the world's children when we reflect on and evaluate our options for action.

2. To be open to study the true values based on the natural law that sustain every human culture.

3. To use science and technology in a full and constructive way, while recognizing that the findings of science have always to be evaluated in the light of the centrality of the human person, of the common good, and of the inner purpose of creation. Science may help us to correct the mistakes of the past, in order to enhance the spiritual and material well-being of the present and future generations. It is love for our children that will show us the path that we must follow into the future.

4. To be humble regarding the idea of ownership and to be open to the demands of solidarity. Our mortality and our weakness of judgment together warn us not to take irreversible actions with what we choose to regard as our property during our brief stay on this earth. We have not been entrusted with unlimited power over creation, we are only stewards of the common heritage.

5. To acknowledge the diversity of situations and responsibilities in the work for a better world environment. We do not expect every person and every

institution to assume the same burden. Everyone has a part to play, but for the demands of justice and charity to be respected the most affluent societies must carry the greater burden, and from them is demanded a sacrifice greater than can be offered by the poor. Religions, governments, and institutions are faced by many different situations; but on the basis of the principle of subsidiarity all of them can take on some tasks, some part of the shared effort.

6. To promote a peaceful approach to disagreement about how to live on this earth, about how to share it and use it, about what to change and what to leave unchanged. It is not our desire to evade controversy about the environment, for we trust in the capacity of human reason and the path of dialogue to reach agreement. We commit ourselves to respect the views of all who disagree with us, seeking solutions through open exchange, without resorting to oppression and domination.

It is not too late. God's world has incredible healing powers. Within a single generation, we could steer the earth toward our children's future. Let that generation start now, with God's help and blessing.

Pope John Paul II
Patriarch Bartholomew I of Constantinople
Rome — Venice,
10 June 2002

Appendix 3

Statement of
H. E. Archbishop Renato R. Martino
Head of the Holy See Delegation
to the United Nations Conference
on Environment and Development
Rio de Janeiro, Brazil,
4 June 1992

The people of the whole world look with keen interest and great expectations to this United Nations Conference on Environment and Development. The challenge facing the international community is how to reconcile the imperative duty of the protection of the environment with the basic right of all people to development.

I. The centrality of the human person.

The Catholic Church approaches both the care and protection of the environment and all questions regarding development from the point of view of the human person. It is the conviction of the Holy See, therefore, that all ecological programmes and all developmental initiatives must respect the full dignity and freedom of whomever might be affected by such programmes. They must be seen in relation to the needs of actual men and women, their families, their values, their unique social and cultural heritage, their responsibility toward future generations. For the ultimate purpose of environmental and developmental programmes is to enhance the quality of human life, to place creation in the fullest way possible at the service of the human family.

The ultimate determining factor is the human person. It is not simply science and technology, nor the increasing means of economic and material development, but the human person, and especially groups of persons, communities and nations, freely choosing to face the problems together, who will, under God, determine the future.[1]

The word environment itself means 'that which surrounds'. This very definition postulates the existence of a centre around which the environment exists. That centre is the human being, the only creature in this world who is not only capable of being conscious of itself and of its surroundings, but is gifted with the intelligence to explore, the sagacity to utilize, and is ultimately responsible for its choices and the consequences of those choices. The praiseworthy heightened awareness of the present generation for all components of the environment, and the consequent efforts at preserving and protecting them, rather than weakening the central position of the human being, accentuate its role and responsibilities.

Likewise, it cannot be forgotten that the true purpose of every economic, social and political system and of every model of development is the integral advancement of the human person. Development is clearly something much more extensive than merely economic progress measured in terms of gross national product. True development takes as its criterion the human person with all the needs, just expectations and fundamental rights that are his or hers.[2]

Complementing respect for the human person and human life is the responsibility to respect all creation. God is creator and planner of the entire universe. The universe and life in all its forms are a testimony to God's creative power, His Love, His enduring presence. All creation reminds us of the mystery and love of God. As the Book of Genesis tells us: 'And God saw everything that He had made, and behold, it was very good.' (Gn 1:31)

II. The moral dimension.

In the very early stages that led to the convening of this Conference, the General Assembly emphasized that 'in view of the global character of major environmental problems, there is a common interest of all countries in pursuing policies aimed at achieving a sustainable and environmentally sound development within a sound ecological balance.'[3]

The Holy See has been and continues to be keenly interested in the issues which this Conference is addressing. During the laborious preparatory phases, the Holy See delegation has carefully and respectfully examined the many proposals of technological, scientific and political nature put forth and appreciates the contributions made by so many participants in the process. Faithful to its nature and its mission, the Holy See has continued to emphasize the rights and the duties, the well-being and the responsibilities of individuals and of societies. For the Holy See the problems of environment and development are, at their root, issues of a moral, ethical nature, from which derive two obligations: the urgent imperative to find solutions and the inescapable demand that every proposed solution meet the criteria of truth and justice.

'Theology, philosophy and science all speak of a harmonious universe, of a "cosmos" endowed with its own integrity, its own internal, dynamic balance. This order must be respected. The human race is called to explore this order, to examine it with due care and to make use of it while safeguarding its integrity.'[4] The Creator has placed the human beings at the centre of creation, making them the responsible stewards, not the exploiting despots, of the world around them. 'On the other hand, the earth is ultimately a common heritage, the fruits of which are for the benefits of all. This has direct consequences for the problem at hand. It is manifestly unjust that a privileged few should continue to accumulate excess goods, squandering available resources, while masses of people are living in conditions of misery at the very lowest level of subsistence. Today, the dramatic threat of ecological breakdown is teaching us the extent to which greed and selfishness -- both individual and collective -- are contrary to the order of creation, an order which is characterized by mutual interdependence.'[5]

III. The resulting obligations: Stewardship and Solidarity.

The concepts of an ordered universe and a common heritage both point to the necessity of developing in the heart of every individual and in the activities of every society a true sense of stewardship and of solidarity.

It is the obligation of a responsible steward to be one who cares for the goods entrusted to him and not one who plunders, to be one who conserves and enhances and not one who destroys and dissipates. Humility, and not arrogance, must be the proper attitude of humankind vis-à-vis the environment. The exciting scientific discoveries of our century have enabled the human mind to pierce with equal success into the infinitesimally small as well as into the immeasurably large. The results have been ambivalent, for we have witnessed that, without ethics, science and technology can be employed to kill as well as to save lives, to manipulate as well as to nurture, to destroy as well as to build.

Responsible stewardship demands a consideration for the common good: no one person, no one group of people in isolation are allowed to determine their relationship with the universe. The universal common good transcends all private interests, all national boundaries, and reaches, beyond the present moment, to the future generations.

Hence, solidarity becomes an urgent moral imperative. We are all part of God's creation -- we live as a human family. The whole of creation is everyone's heritage. All equally created by God, called to share the goods and the beauty of the one world, human beings are called to enter into a solidarity of universal dimensions, 'a cosmic fraternity' animated by the very love that flows from God. Education to solidarity is an urgent necessity of our day. We must learn again to live in harmony, not only with God and with one another, but with creation itself. The 'Canticle of Creatures' of Francis of Assisi could well become the anthem of a new generation that loves and respects in one embrace the Creator and all God's creatures.

Responsible stewardship and genuine solidarity are not only directed to the protection of the environment, but, equally so, to the inalienable right and duty of all peoples to development. The earth's resources and the means to their access and use must be wisely monitored and justly shared. The demands for the care and protection of the environment cannot be used to obstruct the right to development, nor can development be invoked in thwarting the environment. The task of achieving a just balance is today's challenge.

The scandalous patterns of consumption and waste of all kinds of resources by a few must be corrected, in order to ensure justice and sustainable development to all, everywhere in the world. Pope John Paul II has reminded that: 'Simplicity, moderation and discipline, as well as a spirit of sacrifice, must become part of everyday life, lest all suffer the negative consequences of the careless habits of a few.'[6] The developing countries, in their legitimate ambition to improve their status and emulate existing patterns of development, will realize and counteract the danger that can derive to their people and to the world by the adoption of highly wasteful growth strategies hitherto widely employed, that have led humanity into the present situation.

New resources, the discovery of substitute new materials, determined efforts at conservation and recycling programmes have assisted in the protection of known reserves; the development of new technologies has the promise of using resources more efficiently.

For developing nations, at times rich in natural resources, the acquisition and use of new technologies is a clear necessity. Only an equitable global sharing of technology will make possible the process of sustainable development.

When considering the problems of environment and development one must also pay due attention to the complex issue of population. The position of the Holy See regarding procreation is frequently misinterpreted. The Catholic Church does not propose procreation at any cost. It keeps on insisting that the transmission of, and the caring for human life must be exercised with an utmost sense of responsibility. It restates its constant position that human life is sacred; that the aim of public policy is to enhance the welfare of families; that it is the right of the spouses to decide on the size of the family and spacing of births, without pressures from governments or organizations. This decision must fully respect the moral order established by God, taking into account the couple's responsibilities toward each other, the children they already have and the society to which they belong.[7] What the Church opposes is the imposition of demographic policies and the promotion of methods for limiting births which are contrary to the objective moral order and to the liberty, dignity and conscience of the human being. At the same time, the Holy See does not consider people as mere numbers, or only on economic terms.[8] It emphatically states its concern that the poor not be singled out as if, by their very existence, they were the cause, rather than the victims, of the lack of development and of environmental degradation.

Serious as the problem of interrelation among environment, development and population is, it cannot be solved in an over-simplistic manner and many of the most alarming predictions have proven false and have been discredited by a number of recent studies. 'People are born not only with mouths that need to be fed, but also with hands that can produce, and minds that can create and innovate.'[9] As for the environment, just to mention one instance, countries with as few as 5% of the world population are responsible for more than one quarter of the principal greenhouse gas, while countries with up to a quarter of the world population contribute as little as 5% of the same greenhouse gas.

A serious and concerted effort aimed at protecting the environment and at promoting development will not be possible without directly addressing the structural forms of poverty that exist throughout the world. Environment is devastated and development thwarted by the outbreak of wars, when internal conflicts destroy homes, fields and factories, when intolerable circumstances force millions of people to desperately seek refuge away from their lands, when minorities are oppressed, when the rights of the most vulnerable – women, children, the aged and the infirm – are neglected or abused. 'The poor, to whom the earth is entrusted no less than to others, must be enabled to find a way out of their poverty. This will require a courageous reform of structures, as well as new ways of relating among peoples and States.'[10]

Finally, the Holy See invites the international community to discover and affirm that there is a spiritual dimension to the issues at hand. Human beings have the need for and the right to more than clean air and water, to more than economic and technological progress. Human beings are also fragile and an alarm must be sounded against the poisoning of the minds and the corruption of the hearts, both in the developed and developing worlds. The dissemination of hatred, of falsehood and vice, the traffic and use of narcotic drugs, the ruthless self-centeredness which disregards the rights of others -- even the right to life -- are all phenomena that cannot be gauged by technical instrument, but whose chain-effects destroy individuals and societies. Let us strive to give to every man, woman and child a safe and healthy physical environment, let us join forces in providing them with real opportunities for development, but, in the process, let us not allow them to be robbed of their souls. On a closely related level, the aesthetic value of the environment must also be considered and protected, thus adding beauty and inspiring artistic expression to the developmental activities.

The Holy See regards this Conference as a major challenge and a unique opportunity that the people of the world are presenting to the international community. The problems facing today's world are serious indeed and even threatening. Nonetheless, the opportunity is at hand. Avoiding confrontation, and engaging in honest dialogue and sincere solidarity, all forces must be joined in a positive adventure of unprecedented magnitude and cooperation that will restore hope to the human family and renew the face of the earth.

Notes

1   Cf. Pope John Paul II, *Address of to the United Nations Centre for the Environment*, Nairobi, 18 August 1985.

2   Cf. Pope John Paul II, *Address to the 21ˢᵗ Session of the Conference of the Food and Agricultural Organization*, 13 November 1981.

3   Resolution 43/196 of the UN General Assembly, 20 December 1988.

4   Pope John Paul II, Message for the 1990 World Day of Peace, *Peace with God the Creator, Peace with All of Creation*, 8.

5   *Ibid.*.

6   *Ibid.*, 13.

7   Cf. Cardinal Maurice Roy, President of the Pontifical Commission 'Justice and Peace', *Message to U Thant, Secretary-General of the United Nations*, on the occasion of the launching of the Second Development Decade, 19 November 1970.

8   Cf. Pope John Paul II, *Address to Mr. Rafael Salas, Secretary-*

General of the 1984 International Conference on Population, 7 June 1984.

9   Prince Malthus, 'Review and Outlook' in *Wall Street Journal*, 28 April 1992.

10   Pope John Paul II, *Peace with God the Creator, Peace with All of Creation*, 11.

Appendix 4

Bishop Giampaolo Crepaldi, Secretary of the Pontifical Council for Justice and Peace, summarized in an interpretative decalogue the teachings in the tenth chapter of the Compendium of the Social Doctrine of the Church.

1. The Bible lays out the fundamental moral principles of how to affront the ecological question. The human person, made in God's image, is superior to all other earthly creatures, which should in turn be used responsibly. Christ's incarnation and his teachings testify to the value of nature: Nothing that exists in this world is outside the divine plan of creation and redemption.

2. The social teaching of the Church recalls two fundamental points. We should not reduce nature to a mere instrument to be manipulated and exploited. Nor should we make nature an absolute value, or put it above the dignity of the human person.

3. The question of the environment entails the whole planet, as it is a collective good. Our responsibility toward ecology extends to future generations.

4. It is necessary to confirm both the primacy of ethics and the rights of man over technology, thus preserving human dignity. The central point of reference for all scientific and technical applications must be respect for

the human person, who in turn should treat the other created beings with respect.

5.   Nature must not be regarded as a reality that is divine in itself; therefore, it is not removed from human action. It is, rather, a gift offered by our Creator to the human community, confided to human intelligence and moral responsibility. It follows, then, that it is not illicit to modify the ecosystem, so long as this is done within the context of a respect for its order and beauty, and taking into consideration the utility of every creature.

6.   Ecological questions highlight the need to achieve a greater harmony both between measures designed to foment economic development and those directed to preserving the ecology, and between national and international policies. Economic development, moreover, needs to take into consideration the integrity and rhythm of nature, because natural resources are limited. And all economic activity that uses natural resources should also include the costs of safeguarding the environment into the calculations of the overall costs of its activity.

7.   Concern for the environment means that we should actively work for the integral development of the poorest regions. The goods of this world have been created by God to be wisely used by all. These goods should be shared, in a just and charitable manner. The principle of the universal destiny of goods offers a fundamental orientation to deal with the complex relationship between ecology and poverty.

8. Collaboration, by means of worldwide agreements, backed up by international law, is necessary to protect the environment. Responsibility toward the environment needs to be implemented in an adequate way at the juridical level. These laws and agreements should be guided by the demands of the common good.

9. Lifestyles should be oriented according to the principles of sobriety, temperance and self-discipline, both at the personal and social levels. People need to escape from the consumer mentality and promote methods of production that respect the created order, as well as satisfying the basic needs of all. This change of lifestyle would be helped by a greater awareness of the interdependence between all the inhabitants of the earth.

10. A spiritual response must be given to environmental questions, inspired by the conviction that creation is a gift that God has placed in the hands of mankind, to be used responsibly and with loving care. People's fundamental orientation toward the created world should be one of gratitude and thankfulness. The world, in fact, leads people back to the mystery of God who has created it and continues to sustain it. If God is forgotten, nature is emptied of its deepest meaning and left impoverished.

Appendix 5

Cornwall Declaration on Environmental Stewardship

The Cornwall Declaration on Environmental Stewardship is an interfaith statement that expresses the common concerns, beliefs, and aspirations about environmental stewardship from a truly ecumenical Jewish, Catholic, and Protestant perspective. The declaration grew out a meeting of theologians, economists, and scientists held in West Cornwall, Connecticut in October 1999. A final version of the declaration was agreed upon on 1 February 2000. The group established themselves as the *Interfaith Council for Environmental Stewardship.*

Text of the Declaration

The past millennium brought unprecedented improvements in human health, nutrition, and life expectancy, especially among those most blessed by political and economic liberty and advances in science and technology. At the dawn of a new millennium, the opportunity exists to build on these advances and to extend them to more of the earth's people.

At the same time, many are concerned that liberty, science, and technology are more a threat to the environment than a blessing to humanity and nature. Out of shared reverence for God and His creation and love for our neighbors, we Jews, Catholics, and Protestants, speaking for ourselves and not officially on behalf of our respective communities, joined by others of good will, and committed to justice and compassion, unite in this declaration of our common concerns, beliefs, and aspirations.

Our Concerns

Human understanding and control of natural processes empower people not only to improve the human condition but also to do great harm to each other, to the earth, and to other creatures. As concerns about the environment have grown in recent decades, the moral necessity of ecological stewardship has become increasingly clear. At the same time, however, certain misconceptions about nature and science, coupled with erroneous theological and anthropological positions, impede the advancement of a sound environmental ethic. In the midst of controversy over such matters, it is critically important to remember that while passion may energize environmental activism, it is reason – including sound theology and sound science – that must guide the decision-making process. We identify three areas of common misunderstanding:

1. Many people mistakenly view humans as principally consumers and polluters rather than producers and stewards. Consequently, they ignore our potential, as bearers of God's image, to add to the earth's abundance. The increasing realization of this potential has enabled people in societies blessed with an advanced

economy not only to reduce pollution, while producing more of the goods and services responsible for the great improvements in the human condition, but also to alleviate the negative effects of much past pollution. A clean environment is a costly good; consequently, growing affluence, technological innovation, and the application of human and material capital are integral to environmental improvement. The tendency among some to oppose economic progress in the name of environmental stewardship is often sadly self-defeating.

2. Many people believe that 'nature knows best', or that the earth–untouched by human hands–is the ideal. Such romanticism leads some to deify nature or oppose human dominion over creation. Our position, informed by revelation and confirmed by reason and experience, views human stewardship that unlocks the potential in creation for all the earth's inhabitants as good. Humanity alone of all the created order is capable of developing other resources and can thus enrich creation, so it can properly be said that the human person is the most valuable resource on earth. Human life, therefore, must be cherished and allowed to flourish. The alternative–denying the possibility of beneficial human management of the earth–removes all rationale for environmental stewardship.

3. While some environmental concerns are well founded and serious, others are without foundation or greatly exaggerated. Some well-founded concerns focus on human health problems in the developing world arising from inadequate sanitation, widespread use of

primitive biomass fuels like wood and dung, and primitive agricultural, industrial, and commercial practices; distorted resource consumption patterns driven by perverse economic incentives; and improper disposal of nuclear and other hazardous wastes in nations lacking adequate regulatory and legal safeguards. Some unfounded or undue concerns include fears of destructive manmade global warming, overpopulation, and rampant species loss. The real and merely alleged problems differ in the following ways:

a. The former are proven and well understood, while the latter tend to be speculative.

b. The former are often localized, while the latter are said to be global and cataclysmic in scope.

c. The former are of concern to people in developing nations especially, while the latter are of concern mainly to environmentalists in wealthy nations.

d. The former are of high and firmly established risk to human life and health, while the latter are of very low and largely hypothetical risk.

e. Solutions proposed to the former are cost effective and maintain proven benefit, while solutions to the latter are unjustifiably costly and of dubious benefit.

Public policies to combat exaggerated risks can dangerously delay or reverse the economic development necessary to improve not only human life but also human stewardship of the environment. The poor, who are most often citizens of developing nations, are often forced to suffer longer in poverty with its attendant high rates of malnutrition, disease, and mortality; as a consequence, they are often the most injured by such misguided, though well-intended, policies.

Our Beliefs

Our common Judeo-Christian heritage teaches that the following theological and anthropological principles are the foundation of environmental stewardship:

1. God, the Creator of all things, rules over all and deserves our worship and adoration.

2. The earth, and with it all the cosmos, reveals its Creator's wisdom and is sustained and governed by His power and loving kindness.

3. Men and women were created in the image of God, given a privileged place among creatures, and commanded to exercise stewardship over the earth. Human persons are moral agents for whom freedom is an essential condition of responsible action. Sound environmental stewardship must attend both to the demands of human well being and to a divine call for human beings to exercise caring dominion over the earth. It affirms that human well being and the integrity of creation are not only compatible but also dynamically interdependent realities.

4. God's Law—summarized in the Decalogue and the two Great Commandments (to love God and neighbor), which are written on the human heart, thus revealing His own righteous character to the human person—represents God's design for shalom, or peace, and is the

supreme rule of all conduct, for which personal or social prejudices must not be substituted.

5. By disobeying God's Law, humankind brought on itself moral and physical corruption as well as divine condemnation in the form of a curse on the earth. Since the fall into sin people have often ignored their Creator, harmed their neighbors, and defiled the good creation.

6. God in His mercy has not abandoned sinful people or the created order but has acted throughout history to restore men and women to fellowship with Him and through their stewardship to enhance the beauty and fertility of the earth.

7. Human beings are called to be fruitful, to bring forth good things from the earth, to join with God in making provision for our temporal well being, and to enhance the beauty and fruitfulness of the rest of the earth. Our call to fruitfulness, therefore, is not contrary to but mutually complementary with our call to steward God's gifts. This call implies a serious commitment to fostering the intellectual, moral, and religious habits and practices needed for free economies and genuine care for the environment.

Our Aspirations

In light of these beliefs and concerns, we declare the following principled aspirations:

1. We aspire to a world in which human beings care wisely and humbly for all creatures, first and foremost for their fellow human beings, recognizing their proper place in the created order.

2. We aspire to a world in which objective moral principles–not personal prejudices–guide moral action.

3. We aspire to a world in which right reason (including sound theology and the careful use of scientific methods) guides the stewardship of human and ecological relationships.

4. We aspire to a world in which liberty as a condition of moral action is preferred over government-initiated management of the environment as a means to common goals.

5. We aspire to a world in which the relationships between stewardship and private property are fully appreciated, allowing people's natural incentive to care for their own property to reduce the need for collective ownership and control of resources and enterprises, and in which collective action, when deemed necessary, takes place at the most local level possible.

6. We aspire to a world in which widespread economic freedom–which is integral to private, market economies–makes sound ecological stewardship available to ever greater numbers.

7. We aspire to a world in which advancements in agriculture, industry, and commerce not only minimize pollution and transform most waste products into efficiently used resources but also improve the material conditions of life for people everywhere.

Appendix 6

Universal Declaration of Animal Rights

The Universal Declaration of Animal Rights was solemnly proclaimed in Paris on 15 October 1978 at the UNESCO headquarters. The text, revised by the International League of Animal Rights in 1989, was submitted to the UNESCO Director General in 1990 and made public that same year.

Preamble:

Considering that Life is one, all living beings having a common origin and having diversified in the course of the evolution of the species,

Considering that all living beings possess natural rights, and that any animal with a nervous system has specific rights,

Considering that the contempt for, and even the simple ignorance of, these natural rights, cause serious damage to Nature and lead men to commit crimes against animals,

Considering that the coexistence of species implies a recognition by the human species of the right of other animal species to live,

Considering that the respect of animals by humans is inseparable from the respect of men for each other,

It is hereby proclaimed that:

Article 1

All animals have equal rights to exist within the context of biological equilibrium. This equality of rights does not overshadow the diversity of species and of individuals.

Article 2

All animal life has the right to be respected.

Article 3

1. Animals must not be subjected to bad treatments or to cruel acts.

2. If it is necessary to kill an animal, it must be instantaneous, painless and cause no apprehension.

3. A dead animal must be treated with decency.

Article 4

1. Wild animals have the right to live and to reproduce in freedom in their own natural environment.

2. The prolonged deprivation of the freedom of wild animals, hunting and fishing practised as a pastime, as well as any use of wild animals for reasons that are not vital, are contrary to this fundamental right.

Article 5

1. Any animal which is dependent on man has the right to proper sustenance and care.

2. It must under no circumstances be abandoned or killed unjustifiably.

3. All forms of breeding and uses of the animal must respect the physiology and behaviour specific to the species.

4. Exhibitions, shows and films involving animals must also respect their dignity and must not include any violence whatsoever.

Article 6

1. Experiments on animals entailing physical or psychological suffering violate the rights of animals.

2. Replacement methods must be developed and systematically implemented.

Article 7

Any act unnecessary involving the death of an animal, and any decision leading to such an act, constitute a crime against life.

Article 8

1. Any act compromising the survival of a wild species and any decision leading to such an act are tantamount to genocide, that is to say, a crime against the species.

2. The massacre of wild animals, and the pollution and destruction of biotopes are acts of genocide.

Article 9

1. The specific legal status of animals and their rights must be recognised by law.

2. The protection and safety of animals must be represented at the level of Governmental organizations.

Article 10

Educational and schooling authorities must ensure that citizens learn from childhood to observe, understand and respect animals.

Bibliography

Pope JOHN PAUL II, *Peace with God the Creator, Peace with All of Creation* (Message for the Celebration of the World Day of Peace, 1 January 1990).

AA.VV., *Energy for Survival and Development Scripta Varia 57* (Vatican City: Pontifical Academy of Sciences, 1986).

AA.VV., A *Modern Approach to the Protection of the Environment Scripta Varia 75* (Vatican City: Pontifical Academy of Sciences, 1988).

AA.VV., *Man and his environment. Tropical Forests and the Conservation of Species Scripta Varia 84* (Vatican City: Pontifical Academy of Sciences, 1994).

AA.VV., *La questione ecologica* (Roma: A.V.E., 1989).

AA.VV., *La responsabilità ecologica* (Roma: Studium, 1991).

AA.VV., *Ambiente e Tradizione Cristiana* (Brescia: Morcelliana, 1990).

AA.VV., *Christianity and Ecology* (London: Cassell, 1992).

AA.VV., *Priests and People 9/2* (February 1995).

ALLCHIN, A. M., *Wholeness and Transfiguration Illustrated in the Lives of St. Francis of Assisi and St. Seraphim of Sarov.* (London: Oxford University Press, 1974) .

AUER, A., *Etica dell'ambiente* (Brescia: Queriniana, 1988).

IDEM, *Umweltethik. Ein theologischer Beitrag zur ökologischen Diskussion* (Düsseldorf: Patmos Verlag, 1984, 1985).

BAILEY, R. (ed.), *The True State of the Planet* (New York: The Free Press, 1995).

BARBOUR, I. G., *Ethics in an Age of Technology: The Gifford Lectures*, (San Francisco: Harper and Row, 1993)

BELTRÃO, P. C., *Ecologia umana e valori etico–religiosi* (Roma: Editrice Pontificia Università Gregoriana, 1985).

BERNSTEIN, E., *The Splendor of Creation: A Biblical Ecology* (Cleveland, OH: Pilgrim Press, 2005).

BIANCHI, E., *Le ragioni cristiane dell'ecologia* (Treviso: Editrice San Liberale, 2003).

BOUMA-PREDIGER, S., *For the Beauty of the Earth: A Christian Vision for Creation Care* (Grand Rapids, MI: Baker, 2001).

BREMBATI, F., *La tutela dell'ambiente nell'ordinamento comunitario* (Roma: Università degli Studi di Roma *La Sapienza*, 1980).

BROWN, E., *Our Father's World: Mobilizing the Church to Care for Creation* (Downers Grove, IL: InterVarsity Press, 2006).

CAPRIOLI, A. & VACCARO, L. (editori), *Questione ecologica e coscienza cristiana* (Brescia: Morcelliana, 1988).

CARLO, G. L. & SCHRAM, M., *Cell Phones: Invisible Hazards in the Wireless Age: An Insider's Alarming Discoveries About Cancer and Genetic Damage* (New York: Carroll & Graf, 2001).

CASCIOLI, R. & GASPARI, A., *Che tempo farà* (Milano: Piemme, 2008).

COBB, J. B. JR., *Is It Too Late? A Theology of Ecology* (Beverley Hills, CA: Bruce, 1972).

CUMBEY, C., *The Hidden Dangers of the Rainbow: The New Age Movement and Our Coming Age of Barbarism* (Shreveport, La: Huntington House, 1983)

DAMIEN, M., *Gli animali, l'uomo e Dio* (Casale Monferrato: Piemme, 1987).

IDEM, *L'animal, l'homme e Dieu* (Paris: Cerf, 1978).

DE MARZI, M., *San Francesco d'Assisi e l'ecologia* (Roma: Pontificio Istituto Antoniano, 1981).

ENGLAND, R., *The Unicorn in the Sanctuary. The Impact of the New Age on the Catholic Church* (Manasses, Va.: Trinity Communications, 1990).

FARICY, R., *Wind and sea obey Him* (London: SCM, 1982).

FEDERATION OF ASIAN BISHOPS CONFERENCES, «Love for creation: An Asian Response to the Ecological Crisis.» *Catholic International* 4/6 (June 1993) pp.269–272.

GARGANTINI, M., *I Papi e la scienza* (Milano: Jaca, 1985) pp.233–241.

GASPARI, A., *Profeti di Sventura. No Grazie!* (Milano: 21mo Secolo, 1997).

GUROIAN, V., 'Toward ecology as an ecclesial event: Orthodox theology and ecological ethics' in *Communio* 18/1 (Spring 1991), pp. 89–110.

HAFFNER, P., *Mystery of Creation* (Leominster: Gracewing, 1995).

HELLWIG, M. K., *Guests of God: Stewards of Divine Creation* (New York: Paulist Press, 200o).

HODGSON, P. E., *Our Nuclear Future* (Belfast: Christian Journals Ltd., 1983).

HOUGH, A., *God is not 'Green'. A Re–examination of Eco–theology* (Leominster: Gracewing, 1997).

JAKI, S. L., *The Savior of Science* (Washington, D.C.: Regnery Gateway, 1988).

IDEM, *Patterns or Principles and Other Essays* (Bryn Mawr: Intercollegiate Studies Institute, 1995).

IDEM, *The Road of Science and the Ways to God* (Port Huron, MI: Real View Books, 2004).

JONAS, H., *Il principio responsabilità* (Einaudi, 1993).

IDEM, *Dalla fede antica all'uomo tecnologico* (Il Mulino, 1991).

JESUIT SOCIAL JUSTICE SECRETARIAT, 'We live in a broken world. Reflections on Ecology' in *Promotio Iustitiae* 70 (April 1999).

KANTONEN, T. A., *A Theology for Christian Stewardship* (Philadelphia: Muhlenberg Press, 1956).

KEENAN, M., *Care for Creation. Human Activity and the Environment* (Vatican City: Pontifical Council for Justice and Peace, 2000).

KESELOPOULOS, A. G., *Man and the Environment : a Study of St. Simeon the New Theologian* (New York: St. Vladimir's Seminary Press, 2001).

LACHANCE, A.J. & CARROLL, J. E. (eds.), *Embracing Earth. Catholic Approaches to Ecology* (Maryknoll: Orbis Books, 1994).

LAWSON, N., *An Appeal to Reason: A Cool Look at Global Warming* (London: Duckworth, 2008).

MACKAY, A. I., *Farming and Gardening in the Bible* (Emmaus, PA: Rodale Press, 1950).

MAXIMUS THE CONFESSOR, ST, *On the Cosmic Mystery of Jesus Christ* (New York: St. Vladimir's Seminary Press, 2003).

MOLTMANN, J., *The Future of Creation* (Philadelphia: Fortress Press, 1979).

IDEM, *God in Creation: A New Theology of Creation and the Spirit of God*. (San Francisco: Harper and Row, 1985).

MURPHY, C. M., *At Home on Earth. Foundations for a Catholic Ethic of the Environment* (New York: Crossroad, 1989).

PANTEGHINI, G., *Il Gemito della Creazione* (Padova: Edizioni Messaggeri, 1992).

PETRINI, C., *Bioetica, Ambiente, Rischio.* (Roma: Logos Press, 2002).

PIACENTINI, E., *Ecologia Francescana. Approccio morale al problema ecologico agli albori del terzo millennio* (Roma: Bannò, 2002).

PILLA, A. M., *Reverence and Responsability* (Pastoral Letter on the Environment) in *Catholic International* 2/3 (1991), pp. 116–122.

POLKINGHORNE, J., *Science and Providence: God's Interaction with the World* (Boston: New Science Library, 1989).

PRZEWOZNY, B., *La visione cristiana dell'ambiente. Testi del Magistero Pontificio* (Pisa: Giardini, 1991).

RAY, D.L., *Trashing the Planet* (New York: Harper Collins, 1992).

RUST, E. C., *Nature and Man in Biblical Thought* (London: Lutterworth Press, 1953).

IDEM, *Nature—Garden or Desert? An Essay in Environmental Theology* (Waco, TX: Word Books, 1971).

SCHEFFCZYK, L., *Creation and Providence* (New York: Herder, 1970).

SGRECCIA, E. & FISSO, M.B. (eds.), *Etica dell'ambiente* (Roma: Università Cattolica del Sacro Cuore, 1997).

SORRELL, R., *St. Francis of Assisi: Tradition and Innovation in Western Christian Attitudes Toward the Environment* (New York: Oxford University Press, 1988).

SOUTHGATE, C. ET AL., *God, Humanity and the Cosmos. A Textbook in Science and Religion* (Harrisburg, Pa: Trinity Press International, 1999).

TETTAMANZI, D., *Nuova Bioetica Cristiana* (Casale Monferrato: Piemme, 2000), pp. 468–484.

TODISCO, A., *Breviario di Ecologia* (Milano: Rusconi, 1974).

WARD, B. & DUBOS, R., *Only one earth* (Harmondsworth: Penguin, 1972).

ROBINSON, T., & CHATROW, J., *Saving God's Green Earth: Rediscovering the Church's Responsibility to Environmental Stewardship* (Norcross, GA: Ampelon Publishing: 2006)

WHELAN, R., KIRWAN, J. & HAFFNER, P., *The Cross and the Rainforest* (Grand Rapids: Eerdmans, 1996).

IDEM, *Ecología Humana. Respuesta Cristiana al Ambientalismo Radical.* (Santiago de Chile: Libertad y Desarrolo, 1999).

WILLIAMS, T. D., *Who is my Neighbor? Personalism and the Foundations of Human Rights* (Washington D.C.: The Catholic University of America Press, 2005).

ZIZIOULAS, I., *Il creato come eucaristia. Approccio teologico al problema dell'ecologia* (Magnano: Qiqajon, 1994).

Index

From GRACEWING

By the same author

Mystery of Creation

In practically the only recent book in English to give a global picture of the theology of creation, Paul Haffner explores God's masterpiece, the spiritual and material cosmos, from the angels to man and woman centred in Christ. Mystery of Creation touches many well–worn areas of interest in Christian faith and experience, including creation out of nothing, the beginning of the world, man and woman , and original sin.

Many topics of current concern are treated, like the world of spirits, the evolution of the universe and of life, the problem of evil, and the place of animals. Not only does the book take at new look at scientific, ecological and women's issues, it also shows how this universe is our home and yet is a prelude to the New Creation: the best is yet to come!

ISBN 978 0 85244 316 3

The Sacramental Mystery

The seven sacraments lie at the centre of Christian life and experience, for here God the Holy Trinity touches human lives and hearts. This book is one of the few at the present time to offer a global synthesis of the main themes in the sacramental mystery in which the human and divine, the material and the spiritual realms are intimately intertwined. Paul Haffner outlines how the sacraments are the chief means in the Church through which God's people are reconciled to the Father, through His Son, by the power of the Holy Spirit. The treatise illustrates classical issues like the conditions for the validity and the efficacy of the sacraments, as well as the minister, recipient and effects of these sacred mysteries; it deals with particular topics like the necessity of Baptism, the sacrificial character of the Eucharist, and the nature of Marriage. As he examines each sacrament in turn, the work also explores how new ecumenical questions affect Christian sacramental understanding.

'I warmly commend this work on the subject of sacramental theology.'

Archbishop Csaba Ternyák,
Secretary of the Vatican Congregation for the Clergy

ISBN 978 0 85244 476 4

The Mystery of Reason

The Mystery of Reason investigates the enterprise of human thought searching for God. People have always found stepping–stones to God's existence carved in the world and in the human condition. This book examines the classical proofs of God's existence, and affirms their continued validity. It shows that human thought can connect with God and with other aspects of religious experience. Moreover, it depicts how Christian faith is reasonable, and is neither blind nor naked. Without reason, belief would degenerate into fundamentalism; but without faith, human thought can remain stranded on the reef of its own self–sufficiency. The book closes by proposing that the human mind must be in partnership with the human heart in any quest for God.

'This fine work may be seen as a response to the Papal encyclical Fides et Ratio. It is an exploration of the relation-ship between faith and reason, and in so doing it makes use of a variety of approaches including philosophy, theology and contemplation. It is wholly faithful to the vision of the Church.'

Dr Pravin Thevathasan

ISBN 978 0 85244 538 9

The Mystery of Mary

The Blessed Virgin Mary stands at the heart of the Christian tradition. She holds a unique place in the Church's theology, doctrine and devotion, commensurate with her unique position in human history as the Mother of God. In this book, Paul Haffner offers a clear and structured overview of theology and doctrine concerning Mary, within an historical perspective. He outlines the basic scheme of what constitutes Mariology, set in the context of other forms of theological enquiry, and, working through the contribution of Holy Scripture – in the Old Testament forms of prefiguration and the New Testament witness – he proceeds to examine each of the fundamental doctrines that the Church teaches about Our Lady. From the Immaculate Conception to Mary's continuing Motherhood in the Church as Mediatrix of all graces, the reader will find here a sure and steady guide, faithful to tradition and offering a realist perspective, not reducing the concrete aspects of Mary's gifts and privileges to mere symbols on the one hand, and not confusing doctrine and devotionalism on the other.

The Mystery of Mary, with a foreword by Dom Duarte, Duke of Braganza, was published to celebrate the one hundred and fiftieth anniversary of the definition, by Pope Blessed Pius IX, of the dogma of the Immaculate Conception of Our Lady.

ISBN 978 0 85244 650 8

Mystery of the Church

Mystery of the Church presents a global picture of the main themes of current ecclesiology. First, it deals with the institution of the Church and her essential nature. Subsequently the four hallmarks of the Church are described. Her unity and holiness are guaranteed by the sanctity of Christ her Head despite the sinfulness of her members. The catholicity of the Church is also examined from the perspective of Eastern Christendom. The apostolicity of the Church leads to a description of the Petrine Office. The Church is seen as the instrument of salvation, and her relationship with the State and with science is investigated. Finally the Church is pictured as leading to the Kingdom of God.

ISBN 978 0 85244 133 6

CPSIA information can be obtained
at www.ICGtesting.com
Printed in the USA
BVHW031127191221
624277BV00005B/18

9 780852 443682